RF, Microwave, and Millimeter Wave Devices and Circuits and Their Applications

RF, Microwave, and Millimeter Wave Devices and Circuits and Their Applications

Guest Editor

Reza K. Amineh

MDPI

Basel • Beijing • Wuhan • Barcelona • Belgrade • Novi Sad • Cluj • Manchester

Guest Editor
Reza K. Amineh
Department of Electrical and
Computer Engineering
New York Institute of
Technology
New York, NY
USA

Editorial Office
MDPI AG
Grosspeteranlage 5
4052 Basel, Switzerland

This is a reprint of the Special Issue, published open access by the journal *Electronics* (ISSN 2079-9292), freely accessible at: https://www.mdpi.com/journal/electronics/special_issues/07C6627594.

For citation purposes, cite each article independently as indicated on the article page online and as indicated below:

Lastname, A.A.; Lastname, B.B. Article Title. *Journal Name* **Year**, *Volume Number*, Page Range.

ISBN 978-3-7258-4743-3 (Hbk)
ISBN 978-3-7258-4744-0 (PDF)
https://doi.org/10.3390/books978-3-7258-4744-0

Contents

About the Editor . . . . **vii**

Reza K. Amineh
RF, Microwave, and Millimeter Wave Devices and Circuits and Their Applications
Reprinted from: *Electronics* **2025**, *14*, 2844, https://doi.org/10.3390/electronics14142844 . . . . . **1**

In-Cheol Yoo and Chul-Woo Byeon
A Millimeter-Wave CMOS Down-Conversion Mixer with Transformer-Based Harmonic Suppression
Reprinted from: *Electronics* **2025**, *14*, 943, https://doi.org/10.3390/electronics14050943 . . . . . . **5**

Joon-Hyung Kim and Chul-Woo Byeon
A 60 GHz Power Amplifier with Neutralization Capacitors and Compensation Inductors
Reprinted from: *Electronics* **2024**, *13*, 4276, https://doi.org/10.3390/electronics13214276 . . . . . **18**

Daniele Tripoli, Giorgio Maiellaro, Santi Concetto Pavone and Egidio Ragonese
Interstacked Transformer Quad-Core VCOs
Reprinted from: *Electronics* **2024**, *13*, 927, https://doi.org/10.3390/electronics13050927 . . . . . . **28**

Yue Ma, Qifei Du, Wei Zhang, Cheng Liu and Hao Zhang
A Miniaturized Bandpass Filter with Wideband and High Stopband Rejection Using LTCC Technology
Reprinted from: *Electronics* **2024**, *13*, 166, https://doi.org/10.3390/electronics13010166 . . . . . . **40**

Yingze Wang, Xiaoran Li, Yuanze Wang, Xinghua Wang, Zicheng Liu, Fang Han and Quanwen Qi
Analysis and Optimization Design Scheme of CMOS Ultra-Wideband Reconfigurable Polyphase Filters on Mismatch and Voltage Loss
Reprinted from: *Electronics* **2024**, *13*, 658, https://doi.org/10.3390/electronics13030658 . . . . . . **57**

Seongjin Jang and Changkun Park
Design Techniques for L-C-L T-Type Wideband CMOS Phase Shifter with Suppressed Phase Error
Reprinted from: *Electronics* **2023**, *12*, 4368, https://doi.org/10.3390/electronics12204368 . . . . . **74**

Philip Ayiku Dzagbletey and Jae-Young Chung
Millimeter-Wave Choke Ring Antenna with Broad HPBW and Low Cross-Polarization for 28 GHz Dosimetry Studies
Reprinted from: *Electronics* **2024**, *13*, 3531, https://doi.org/10.3390/electronics13173531 . . . . . **85**

Kejia Zhao, He Chen, Xiangjun Li, Jie Sun, Bo Li, Dexian Yan and Lanlan Li
Time Domain Simulated Characterization of the Coplanar Waveguide in an On-Chip System for Millimeter Waveform Metrology
Reprinted from: *Electronics* **2024**, *13*, 145, https://doi.org/10.3390/electronics13010145 . . . . . . **104**

Vishnu S. Pendyala and Milind Patil
Multi-Link Prediction for mmWave Wireless Communication Systems Using Liquid Time-Constant Networks,Long Short-Term Memory, and Interpretation Using Symbolic Regression
Reprinted from: *Electronics* **2024**, *13*, 2736, https://doi.org/10.3390/electronics13142736 . . . . . **113**

Mona Heydari and Reza K. Amineh
Expedited Near-Field Holographic Microwave Imaging with an Azimuthally Distributed Antenna Array
Reprinted from: *Electronics* **2025**, *14*, 2518, https://doi.org/10.3390/electronics14132518 . . . . . **126**

About the Editor

Reza K. Amineh

Reza K. Amineh is currently an Associate Professor in the Department of Electrical and Computer Engineering at the New York Institute of Technology. Before that, he was a Principal Scientist in the Department of Sensor Physics at Halliburton Co. He received his Ph.D. degree in electrical engineering from McMaster University, Canada, in 2010. He was a Post-Doctoral Fellow at the University of Toronto and McMaster University, from 2012 to 2013 and from 2010 to 2012, respectively. He was a Ph.D. intern with the Advanced Technology Group, BlackBerry, in 2009. He has authored/co-authored over 100 journal and conference papers, three book chapters, and a book titled *Real-Time Three-Dimensional Imaging of Dielectric Bodies Using Microwave/Millimeter Wave Holography*, published by Wiley & IEEE Press. He contributed to more than 30 US patents in applied electromagnetics while working at Halliburton Co. and received several industrial awards. His research interests include applied electromagnetics with applications in imaging and sensing. Amineh was selected for the US Department of Energy's (DOE) Visiting Faculty Program at the Brookhaven National Laboratory (BNL) for the Summer 2024 and Summer 2025 terms. In 2023, he was a recipient of the IEEE Region 1 Technological Innovation (Academic) Award "for developing innovative imaging and sensing systems using electromagnetic waves". In 2022, Amineh received the "Presidential Excellence Award in Student Engagement in Research, Scholarship, or Creative Activities" from New York Tech. He was also a recipient of the prestigious Banting Post-Doctoral Fellowship from the Government of Canada in 2012 and the Ontario Ministry of Research and Innovation (OMRI) Post-Doctoral Fellowship in 2010.

electronics

MDPI

Editorial

RF, Microwave, and Millimeter Wave Devices and Circuits and Their Applications

Reza K. Amineh

Department of Electrical and Computer Engineering, New York Institute of Technology, New York, NY 10023, USA; rkhalaja@nyit.edu

1. Introduction

The recent progress in the development of cost-effective, compact, and highly integrated high-frequency circuits in the RF, microwave, and millimeter-wave domains has significantly broadened the scope of these technologies across both traditional and emerging application areas. While RF and microwave systems were historically confined to military and aerospace use due to their complexity and high cost [1], advances in semiconductor fabrication technologies, such as CMOS and SiGe BiCMOS, have drastically reduced their footprint and cost, allowing their deployment in consumer-grade electronics, automotive systems, healthcare devices, and the burgeoning Internet of Things (IoT) ecosystem [2,3].

This Special Issue of *Electronics*, titled "RF, Microwave, and Millimeter Wave Devices and Circuits and Their Applications", presents a comprehensive overview of recent innovations in the design of high-frequency systems and their multidisciplinary applications. The selected contributions span a wide array of domains—from power amplifiers, mixers, and phase shifters to advanced antenna arrays, metamaterial-inspired filters, etc.—demonstrating how the latest breakthroughs in circuit topology, device architecture, and packaging techniques are enabling high-performance systems at frequencies ranging from a few to tens of gigahertz.

The rapid expansion of wearable sensing and imaging systems in particular, empowered by conformal and miniaturized RF components, is opening up new avenues in personalized healthcare, biomedical diagnostics, and smart textiles [4,5]. Similarly, the increasing availability of the millimeter-wave spectrum (e.g., 28 GHz, 60 GHz) has led to explosive growth in high-data-rate communication systems for 5G and beyond, requiring innovations in front-end circuits, phased arrays, and beamforming architectures [6]. Other topics featured in this issue include wireless power transfer, non-contact sensors for industrial automation, energy-efficient radar systems, precision time-domain metrology, and reconfigurable systems using RF micro-electromechanical systems (MEMS) switches [7,8].

Furthermore, this Special Issue sheds light on the growing role of artificial intelligence and machine learning in the design and adaptation of RF/microwave circuits, such as for real-time link prediction in dynamic mmWave environments. With predictive modeling and symbolic regression, smart RF systems can now operate more efficiently and robustly in complex, multi-link settings. The fusion of circuit design, signal processing, and intelligent control represents a key trend toward self-optimizing and context-aware RF systems.

In summarizing these contributions, this issue not only highlights the breadth of current innovation but also points to critical emerging challenges. These include the need for better integration between analog and digital domains, improved thermal management in mmWave systems, robust packaging for harsh environments, etc. As we stand at the

https://doi.org/10.3390/electronics14142844

convergence of wireless connectivity, sensing, and computation, the opportunities for future research and impactful applications are vast.

We hope that this Special Issue will serve as both a valuable reference and a source of inspiration for engineers, researchers, and industry practitioners working at the frontier of RF, microwave, and millimeter-wave technologies.

2. Summary of the Contributions

We proudly present this compendium of 10 impactful papers that span advanced circuit design, component innovation, and applied systems in high-frequency electronics. The contributions span a wide range of applications, including (1) high-frequency CMOS and mixed-signal ICs, (2) filters, phase shifters, and signal conditioning, (3) antennas and propagation structures, (4) hybrid RF-AI systems, and (5) microwave imaging.

Related to the first topic, Yoo and Byeon present a 60 GHz CMOS down-conversion mixer realized in a 65 nm process. Their design integrates an LO buffer and transconductance stage with a transformer network that actively suppresses second harmonics. The resulting circuit achieves a conversion gain of ~6 dB, a noise figure around 6 dB, strong harmonic suppression (−26 dBc), and high linearity, all while consuming just 7 mW within a compact 0.51 mm^2 die. This positions it as a prime candidate for energy-efficient 5G receiver front-ends. In the second article, Kim and Byeon developed a high-performance 60 GHz CMOS power amplifier featuring both neutralization capacitors and compensation inductors. By mitigating the feedback and frequency-dependent gain roll-off inherent to mmWave PAs, they achieved over 13.4 dBm output power and more than 18 dB of gain, while preserving unconditional stability across the operating band. This advancement offers a robust mmWave transmitter solution suitable for compact wireless platforms. In the third contribution, Tripoli et al. designed a quad-core voltage-controlled oscillator using interstacked transformers to share bias and tank circuitry efficiently. It is an ideal candidate for integration in frequency synthesizers and local oscillators in mmWave communication systems.

In relation to the second topic, Ma et al. proposed a compact Low-Temperature Co-fired Ceramic (LTCC) band-pass filter. Despite its small size, it achieved greater than 44 dB stopband rejection and a smooth wideband response, which makes it well suited for RF front-end modules in IoT, handheld, and wearable devices. In the fifth contribution, Wang et al. explored ultra-wideband polyphase filter designs in CMOS, focusing on mitigating phase mismatches and insertion loss through adaptive biasing techniques. Their measured phase error remained below 0.3° and the amplitude mismatch was within 0.098 dB across the 2 to 8 GHz range, ensuring precise quadrature outputs vital for high-fidelity modulation in communications systems. In the sixth contribution, Jang and Park presented an L-C-L T-network phase shifter optimized for wideband performance in phased-array applications. The design demonstrates phase error under 1° over the 21.5 GHz to 40.0 GHz range, a flat amplitude response, and low insertion loss across its band, enabling precise beam steering with minimal signal degradation.

In relation to the third topic, Dzagbletey and Chung described a 28 GHz choke-ring antenna exhibiting a broad 60° half-power beamwidth and cross-polarization suppression below 28 dB. These properties make it highly suitable for mmWave applications involving precise dosimetry measurements and beamforming calibration in both research and deployment scenarios. In the eighth contribution, Zhao et al. delivered a detailed time-domain study of on-chip coplanar waveguide structures, exploring how sub-micron geometric variations affect amplitude and phase fidelity at mmWave frequencies. Their simulation-based work underscores the importance of fabrication precision for high-performance interconnects and chip-level signal integrity.

In relation to the fourth topic, Pendyala and Patil examined the prediction of mmWave channel quality across multiple simultaneous links using a hybrid of liquid time-constant networks, LSTMs, and symbolic regression. This combination significantly reduces prediction error from ~3.4 dB to ~0.25 dB while providing interpretability to the resulting models, thus offering adaptive insight for beam steering and network adaptation.

Lastly, in the tenth contribution, Heydari and Amineh introduced a groundbreaking near-field microwave imaging system that arranges transmit–receive antenna elements azimuthally around a target volume. This configuration eliminates the need for time-consuming mechanical scanning along the azimuthal direction while achieving a comparable spatial resolution. The authors experimentally demonstrate that by using a 360° antenna distribution, the acquisition time can be reduced by more than 7-fold without sacrificing image clarity, paving the way for real-time biomedical monitoring and rapid non-destructive testing.

3. Conclusions

This Special Issue showcases the progression of high-frequency electronics—from front-end CMOS ICs (mixers, PAs, VCOs) to advanced filtering networks, antennas, AI-enhanced communication systems, and microwave imaging. The research within contributes to advancing performance, miniaturization, and reconfigurability, presenting a clear trajectory toward 5G/6G communications, IoT sensing, biomedical wearables, and high-precision measurement systems.

We thank all the authors for their ingenuity, the reviewers for their rigorous evaluations, and the *Electronics* editorial team for their unwavering support. It is our hope that this issue will serve as a catalyst for future research and technological breakthroughs in the RF, microwave, and millimeter-wave community.

Funding: This article received no external funding.

Conflicts of Interest: The author declares no conflicts of interest.

List of Contributions

1. Yoo, I.-C.; Byeon, C.-W. A Millimeter-Wave CMOS Down-Conversion Mixer with Transformer-Based Harmonic Suppression. *Electronics* **2025**, *14*, 943. https://doi.org/10.3390/electronics14050943
2. Kim, J.-H.; Byeon, C.-W. A 60 GHz Power Amplifier with Neutralization Capacitors and Compensation Inductors. *Electronics* **2024**, *13*, 4276. https://doi.org/10.3390/electronics13214276
3. Tripoli, D.; Maiellaro, G.; Pavone, S.C.; Ragonese, E. Interstacked Transformer Quad-Core VCOs. *Electronics* **2024**, *13*, 927. https://doi.org/10.3390/electronics13050927
4. Ma, Y.; Du, Q.; Zhang, W.; Liu, C.; Zhang, H. A Miniaturized Bandpass Filter with Wideband and High Stopband Rejection Using LTCC Technology. *Electronics* **2024**, *13*, 166. https://doi.org/10.3390/electronics13010166
5. Wang, Y.; Li, X.; Wang, Y.; Wang, X.; Liu, Z.; Han, F.; Qi, Q. Analysis and Optimization Design Scheme of CMOS Ultra-Wideband Reconfigurable Polyphase Filters on Mismatch and Voltage Loss. *Electronics* **2024**, *13*, 658. https://doi.org/10.3390/electronics13030658
6. Jang, S.; Park, C. Design Techniques for L-C-L T-Type Wideband CMOS Phase Shifter with Suppressed Phase Error. *Electronics* **2023**, *12*, 4368. https://doi.org/10.3390/electronics12204368
7. Dzagbletey, P.A.; Chung, J.-Y. Millimeter-Wave Choke Ring Antenna with Broad HPBW and Low Cross-Polarization for 28 GHz Dosimetry Studies. *Electronics* **2024**, *13*, 3531. https://doi.org/10.3390/electronics13173531
8. Zhao, K.; Chen, H.; Li, X.; Sun, J.; Li, B.; Yan, D.; Li, L. Time Domain Simulated Characterization of the Coplanar Waveguide in an On-Chip System for Millimeter Waveform Metrology. *Electronics* **2024**, *13*, 145. https://doi.org/10.3390/electronics13010145

9. Pendyala, V.S.; Patil, M. Multi-Link Prediction for mmWave Wireless Communication Systems Using Liquid Time-Constant Networks, Long Short- Term Memory, and Interpretation Using Symbolic Regression. *Electronics* **2024**, *13*, 2736. https://doi.org/10.3390/electronics13142736
10. Heydari, M.; Amineh, R.K. Expedited Near-Field Holographic Microwave Imaging with an Azimuthally Distributed Antenna Array. *Electronics* **2025**, *14*, 2518. https://doi.org/10.3390/electronics14132518

References

1. Pozar, D.M. *Microwave Engineering*, 4th ed.; Wiley: Hoboken, NJ, USA, 2011.
2. Lee, T.H. *The Design of CMOS Radio-Frequency Integrated Circuits*, 2nd ed.; Cambridge University Press: Cambridge, UK, 2003.
3. Rebeiz, G.M. *RF MEMS: Theory, Design, and Technology*; Wiley: Hoboken, NJ, USA, 2003.
4. Abbasnia, A.; Ravan, M.; Amineh, R.K. Elbow Gesture Recognition with an Array of Inductive Sensors and Machine Learning. *Sensors* **2024**, *24*, 4202. [CrossRef] [PubMed]
5. Liu, J.; Liu, M.; Bai, Y.; Zhang, J.; Liu, H.; Zhu, W. Recent Progress in Flexible Wearable Sensors for Vital Sign Monitoring. *Sensors* **2020**, *20*, 4009. [CrossRef] [PubMed]
6. Rappaport, T.S.; Sun, S.; Mayzus, R.; Zhao, H.; Azar, Y.; Wang, K. Millimeter Wave Mobile Communications for 5G Cellular: It Will Work! *IEEE Access* **2013**, *1*, 335–349. [CrossRef]
7. Deng, Z.; Wang, Y.; Lai, C. Design and Analysis Okasili, I.; Elkhateb, A.; Littler, T. A Review of Wireless Power Transfer Systems for Electric Vehicle Battery Charging with a Focus on Inductive Coupling. *Electronics* **2022**, *11*, 1355.
8. Deng, Z.; Wang, Y.; Lai, C. Design and analysis of pattern reconfigurable antenna based on RF MEMS switches. *Electronics* **2023**, *12*, 3109. [CrossRef]

electronics

MDPI

Article

A Millimeter-Wave CMOS Down-Conversion Mixer with Transformer-Based Harmonic Suppression

In-Cheol Yoo and Chul-Woo Byeon *

School of Electronics and Electrical Engineering, College of Engineering, Dankook University, Yongin-si 16890, Republic of Korea; dlscjf0716@dankook.ac.kr

* Correspondence: cwbyeon@dankook.ac.kr; Tel.: +82-31-8005-3632

Abstract: In this paper, we present a millimeter-wave CMOS down-conversion mixer designed for 5G cellular communications. The proposed mixer integrates a local oscillator buffer, an RF transconductance (Gm) stage, and a switching stage. A transformer-based harmonic suppression technique and separate RF Gm stage and switching stage are employed to achieve a low noise figure (NF), high conversion gain (CG), and effective harmonic suppression. Intermodulation and gain characteristics are analyzed, demonstrating enhanced harmonic suppression, high gain, and low NF. Implemented in 65 nm CMOS technology, the proposed mixer occupies a core chip area of 0.51 mm^2 and consumes a dc power of 7 mW. The implemented design achieves a CG of 6.4 dB, an NF of 6.1 dB, and an output third-order intercept point of 9.0 dBm at an RF frequency of 38.2 GHz. Additionally, harmonic suppression exceeds −26 dBc, highlighting the performance advantages of the proposed architecture.

Keywords: 5G; CMOS; down-conversion mixer; harmonic suppression; millimeter-wave; transformer

1. Introduction

Recently, the standardization of fifth-generation (5G) New Radio (NR) has garnered significant attention for its ability to enable high-speed data transmission and massive connectivity, leveraging millimeter-wave (mm-wave) frequency bands. Frequency Range 2 (FR2), spanning 24.25–71 GHz, includes bands such as 24–30 GHz and 36–42 GHz, which have been allocated to various countries [1–3]. In these frequency bands, phased-array techniques have been proposed to extend communication distances and improve data rates [4–12].

In mm-wave 5G transceivers, the mixer plays a critical role as it facilitates modulation, demodulation, and spectrum shifting. Key performance metrics for down-conversion mixers include high conversion gain (CG), low noise figure (NF), high linearity, effective harmonic suppression, and low power consumption. These parameters are essential to achieving optimal receiver performance in the mm-wave band.

Previous studies have made significant strides in developing wideband, multiband, and high-performance down-conversion mixers for mm-wave applications [13–22]. CMOS technology is widely adopted for its low cost and seamless integration with digital baseband circuits. Many mm-wave CMOS mixers utilize double-balanced Gilbert-cell topologies [15,18,19,22] or separate RF transconductance (Gm) and switching stages [14,16,17,20,21]. For example, a Gilbert-cell mixer with an intermediate frequency (IF) buffer amplifier achieves a high CG of 4.8 dB but suffers from a high NF of 13.5 dB and requires a large local oscillator (LO) input power of 5 dBm [15]. Similarly, mixers with separate switching and Gm stages demonstrate broadband performance using band-switchable capacitors and transformers but exhibit limited CG (<3.1 dB) and high NF (>11 dB) [16,17]. Efforts to enhance performance have introduced mixers with body self-forward bias [18], large-resistance loads [21], or multi-gate transistors [22]. While these designs achieve specific performance gains, they often encounter trade-offs such as high NF, limited CG, or poor LO-RF isolation. For instance, a mixer employing multi-gate transistors (MGTR) achieves exceptional linearity with an IIP3 of 20.87 dBm but exhibits poor CG (−14.4 dB) and NF (16.75 dB) [22].

To address these challenges, this study presents a mm-wave CMOS down-conversion mixer optimized for 5G handheld devices, including smartphones, virtual reality (VR), augmented reality (AR), and gaming consoles. The proposed mixer integrates an LO buffer amplifier, an RF Gm stage, and a switching stage while employing transformer-based harmonic suppression. This architecture achieves a CG of 6.4 dB, an NF of 6.1 dB, a LO-RF isolation exceeding 46 dB, and a 1 dB compression point (IP1dB) of −6 dBm. These results demonstrate that the proposed mixer is well-suited for low-power mm-wave 5G handheld applications.

2. Design Methodology

2.1. Design Considerations

Figure 1 illustrates the block diagram of the proposed down-conversion mixer designed for a super-heterodyne receiver. The mixer converts RF signals (37–41 GHz) to IF signals (10–12 GHz) using an LO frequency range of 27–29 GHz. The design includes a switching stage, an RF Gm stage, an LO buffer amplifier, and transformer-based harmonic suppression.

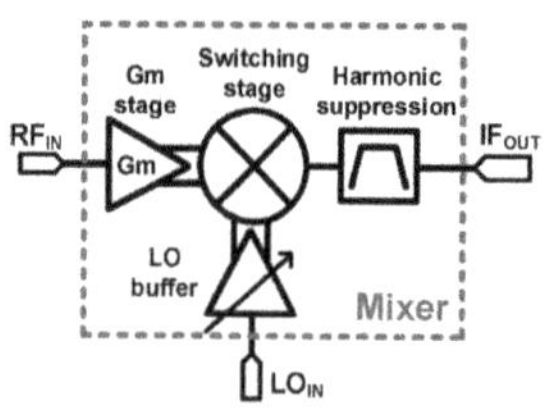

Figure 1. Block diagram of down-conversion mixer.

In conventional Gilbert-cell mixers, large parasitic capacitance at the drain of the Gm stage leads to reduced CG and degraded NF. Additionally, using a single current source for both the Gm stage and switching stage results in performance trade-offs: insufficient current in the Gm stage degrades CG and NF, while excessive current in the switching stage slows down switching speed, further impairing CG and NF. To address these challenges, the proposed mixer separates the Gm stage and switching stage, optimizing gain, NF, and linearity [23].

The LO buffer amplifier is designed with a variable gain function to compensate for input power variations and deliver optimal LO power to the switching stage [24]. Furthermore, in a super-heterodyne receiver, suppressing unwanted harmonics is essential to ensure signal integrity.

Figure 2 presents the schematic of the proposed down-conversion mixer tailored for 5G handheld devices. The design incorporates an RF Gm stage, LO buffer amplifier, switching stage, and transformer-based harmonic suppression to perform RF-to-IF conversion from 39 GHz to 11 GHz. Separating the Gm stage and switching stage enhances NF, CG, and linearity. Additionally, a double-balanced mixer topology is employed to improve

LO rejection compared to single-balanced mixers. Transformer-based harmonic suppression is utilized to effectively reject unwanted harmonics, ensuring robust performance in 5G applications.

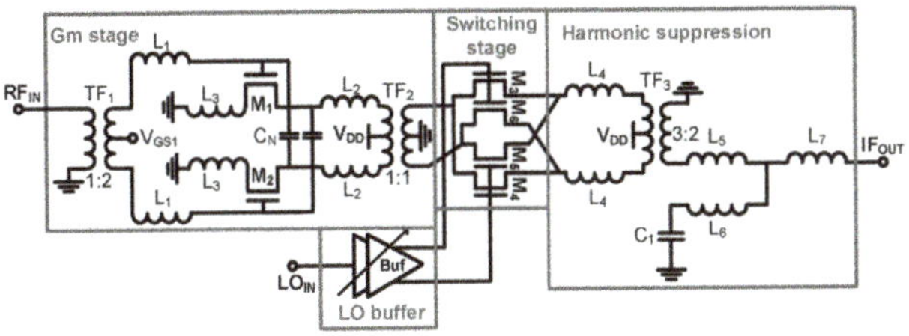

Figure 2. Schematic of proposed down-conversion mixer.

2.2. Transconductance Stage

The Gm stage employs a differential common source configuration with a capacitance neutralization technique [25]. The gate-to-drain parasitic capacitance (C_{gd}) degrades stability, power gain, and reverse isolation due to unwanted feedback. To mitigate this, the capacitance neutralization technique enhances stability, power gain, and LO-to-RF isolation, while improving the mixer's NF by boosting the Gm stage's gain.

The differential RF signal is applied to the switching stage via a differential amplifier comprising transformers (TF_1, TF_2) and a neutralization capacitor (C_N). Transformer, TF_1, along with inductors, L_1 and L_3, achieves simultaneous gain and noise matching, while TF_2 and inductor L_2 provide conjugate matching between the Gm and switching stages.

To optimize the design, a custom metal–oxide–metal (MOM) capacitor [25] replaces the standard foundry-provided MOM capacitor for C_N. This custom MOM capacitor utilizes a multi-stack metal plate design, reducing interconnections and chip area while improving the quality factor. Simulation results show minimal process variation (less than ±2%), ensuring consistent performance.

2.3. Switching Stage with Transformer-Based Harmonic Suppression

The switching stage consists of switching transistors (M_3–M_6) and transformer TF_3 with harmonic suppression. Figure 3 illustrates the three- and two-dimensional layout of the switching stage, designed for symmetry to minimize LO leakage and improve LO-to-RF isolation.

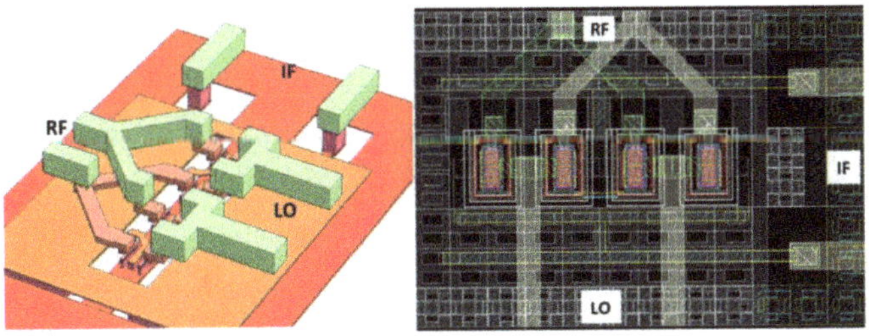

Figure 3. Three- and two-dimensional layout of switching stage.

The layout ensures that the IF output is isolated from RF and LO signal lines using a grounded middle metal layer, while RF and LO signals are routed through the top two metal layers and top metal layer, respectively. This separation and symmetric layout prevent signal interference and enhance performance metrics such as LO leakage and RF-to-LO isolation.

Figure 4a illustrates a transformer model based on [26]. The transformer is represented as an ideal transformer with a turn ratio of 1:n/k, where the primary and secondary leakage inductances are modeled as series components with shunt inductors $(1-k^2)L_P$ and k^2L_P, respectively. Here, L_P is the inductance of the primary winding; R_{P1} and R_{P2} are the parasitic resistances of the primary and secondary windings, respectively; C_{P1} and C_{P2} are the parasitic capacitances, including the winding capacitance; k is the magnetic coupling coefficient, and n is the turn ratio ($\sqrt{L_S/L_P}$), where L_S is the inductance of the secondary winding.

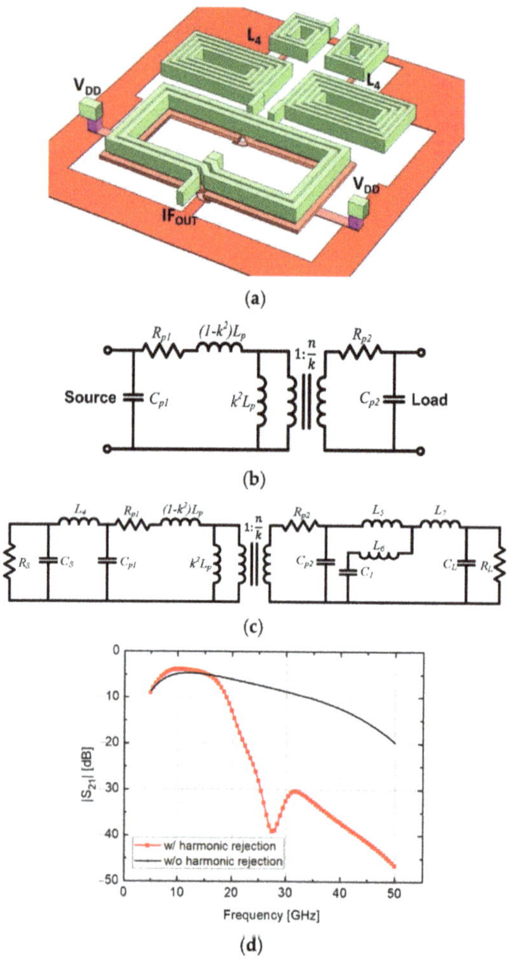

Figure 4. (**a**) 3D layout of the transformer. (**b**) Transformer equivalent circuit model. (**c**) Matching network of the transformer with harmonic suppression. (**d**) Simulated insertion losses of transformers with and without harmonic suppression.

Figure 4a–c depict the 3D layout of the transformer, its equivalent circuit model, and the matching network with harmonic suppression at the switching stage output. The source resistance (R_S) and (C_S) represent the impedance looking into the switching transistors from the drain side, with a source impedance of 88.5-j438.4 Ω at 11 GHz. Similarly, the load resistance (R_L) and capacitance (C_L) represent the impedance looking into the linearization stage, with a load impedance of 26.5-j107.1 Ω at 11 GHz. Incorporating harmonic suppression and the matching network improves the simulated insertion loss by 3.6 dB within the 10–12 GHz range, and the transformer demonstrates effective high-frequency harmonic suppression. Specifically, the insertion loss is measured at 3.9 dB, 6.3 dB, and 38.9 dB at 11 GHz (IF), 17 GHz ($2f_{LO}$-f_{RF}), and 28 GHz (LO frequency), respectively.

The transformer TF_3 is employed for differential-to-single conversion and is constructed using the top two metal layers for magnetic coupling. Its physical primary-to-secondary winding turn ratio is 3:2, with $n = 0.71$ and $k = 0.85$. The parasitic parameters include $R_{S1} = 7.5\ \Omega$, $R_{S2} = 7\ \Omega$, $C_{S1} = 40$ fF, and $C_{S2} = 45$ fF. To further optimize performance, inductors L_4–L_5 form a matching network that reduces insertion loss and rejects high-frequency components. Additionally, a shunt resonance circuit comprising L_6 and C_1 is employed for harmonic suppression and conjugate matching with L_7.

Figure 4d compares the insertion losses of transformers with and without harmonic suppression, highlighting the effectiveness of the proposed design. The primary unwanted signals of the down-conversion mixer are $2f_{LO}$-f_{RF}, $2f_{RF}$-$2f_{LO}$, and f_{LO}. The transformer with harmonic rejection improves suppression by 1 dB, 13 dB, and 31 dB at 17 GHz ($2f_{LO}$-f_{RF}), 22 GHz ($2f_{RF}$-$2f_{LO}$), and 28 GHz (f_{LO}), respectively.

2.4. LO Buffer Amplifier

Figure 5 illustrates the schematic of the proposed LO buffer amplifier [24]. The design incorporates a cascode current-steering variable gain amplifier and a common-source amplifier with a capacitance neutralization technique. The LO buffer amplifier ensures a constant and sufficiently high voltage swing for the mixer core across varying input power levels. To achieve this, it included a gain control function that maintained constant output power across all gain control states.

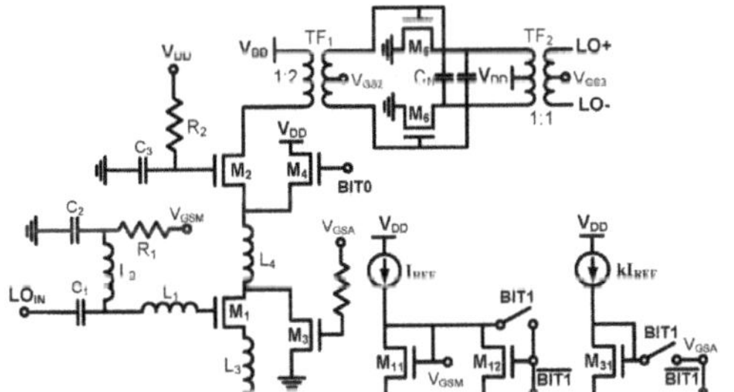

Figure 5. Simplified schematic of proposed LO buffer amplifier.

A conventional current-steering variable gain amplifier [27] adjusts gain by steering the current of the common gate transistor (M_2). However, it typically offers limited output power. To address this limitation and meet linearity requirements, the proposed design employs additional transistors (M_3–M_4) for 2 dB (BIT0) and 4 dB (BIT1) gain control, respectively, as shown in Figure 5. By steering the current through M_1, the design achieves variable gain while maintaining constant output power, driven by the stable current of M_2.

The differential common-source amplifier (M_5–M_6) with capacitance neutralization and transformers (TF_1–TF_2) enhances the LO buffer amplifier's performance by providing high gain, improved linearity, and a well-balanced differential signal. This configuration minimizes gain and phase imbalances, reducing the linearity burden on the first stage.

The simulation results indicate that the proposed LO buffer amplifier achieves a power gain of 15–21 dB in 2 dB steps, with an output 1 dB compression point ranging from −1 dBm to 0.3 dBm. It delivers a 0.7-$V_{P\text{-}P}$ LO signal to the switching stage's gate from an LO input power of −20 dBm. Compared to the conventional current-steering variable gain amplifier, the proposed design demonstrates a 4 dB higher output 1 dB compression point in simulations.

The two bottom metal layers are utilized as a ground plane with a mesh pattern. The inductors, transformers, neutralization capacitors, and interconnections are all modeled and analyzed using the HFSS 3-D EM simulator developed by Ansys, Canonsburg, US. All transistors feature a gate length of 60 nm.

With transformer-based harmonic suppression and the separation of the Gm stage and switching stage, the proposed down-conversion mixer, illustrated in Figure 2, achieves a simulated CG of 5.2–6.4 dB and a NF of less than 7.3 dB over an RF frequency range of 37–41 GHz and an LO frequency range of 27–29 GHz. The mixer demonstrates an input third-order intercept point (IIP3) of 3.1 dBm, an NF of 6.1 dB, a LO-to-RF isolation exceeding 46 dB, and an IP1 dB of −6 dBm.

3. Results

The proposed down-conversion mixer circuit was implemented in a standard 65 nm CMOS technology. Figure 6 shows a chip photograph of the proposed down-conversion mixer. The chip occupies an area of 0.57×0.9 mm^2, excluding the pads. The mixer consumes 18 mW of DC power from a 1 V supply voltage, including the LO buffer amplifier's power consumption of 11 mW. Figure 7 illustrates the measurement setups for S-parameters, CG, NF, and power-handling capability. The mixer's performance was measured on-wafer using ground–signal–ground (GSG) and ground–signal–signal–ground (GSSG) probes. The S-parameters were measured using a Keysight E8361A vector network analyzer. The NF, CG, IP1dB, and IIP3 were measured with a noise figure analyzer, signal generators, and a spectrum analyzer. Losses from the probe tips, adapters, and coaxial cables were de-embedded from the raw measurement data.

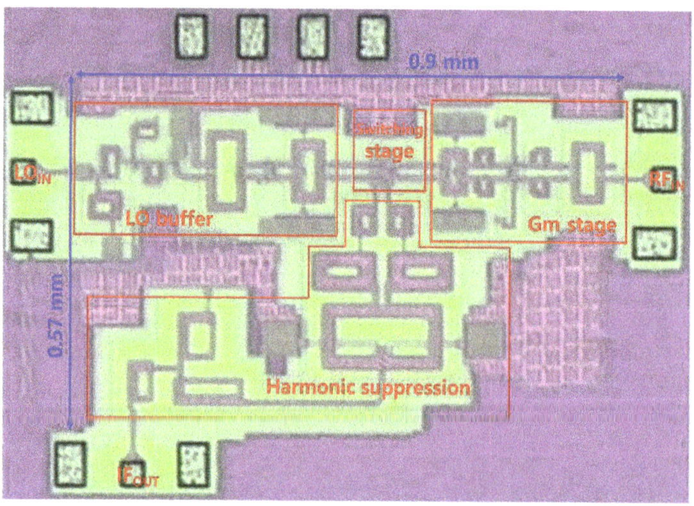

Figure 6. Microphotograph of proposed mixer.

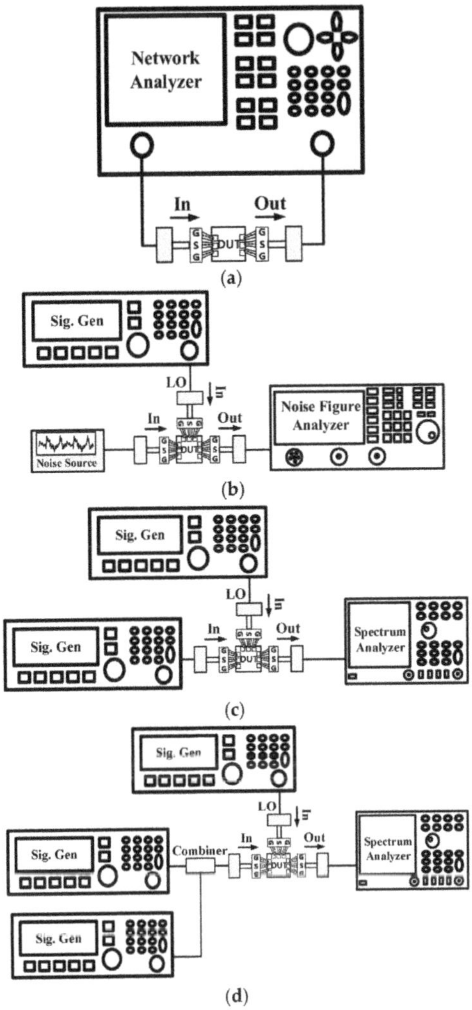

Figure 7. Measurement setups for (**a**) S-parameter, (**b**) NF, (**c**) CG and IP1dB, and (**d**) IIP3.

Figure 8 shows the simulated and measured return losses of the proposed mixer. The measured return losses of the IF and LO ports exceed 10 dB over 8–14 GHz and 26–31 GHz, respectively. The measured return loss of the RF port is better than 8 dB over 36–41 GHz. For the 2-bit gain control in the LO buffer amplifier, the IF and RF ports show consistent return loss performance. Simulated data show good agreement with the measurement results. Figure 9 depicts the simulated and measured NFs of the mixer. The measured single-sideband NF is less than 7.3 dB over an RF frequency range of 37–41 GHz.

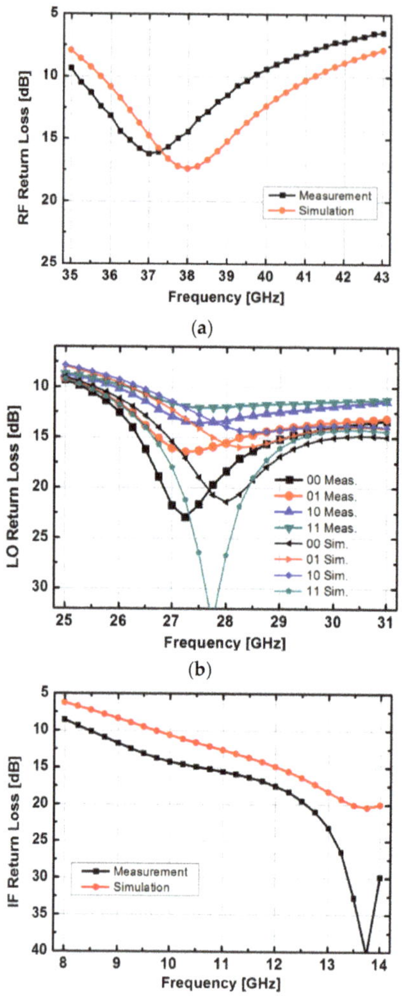

Figure 8. Measured and simulated return losses of the proposed mixer: (**a**) RF port, (**b**) LO port, and (**c**) IF port. For the LO port, return losses of 2-bit gain control are shown.

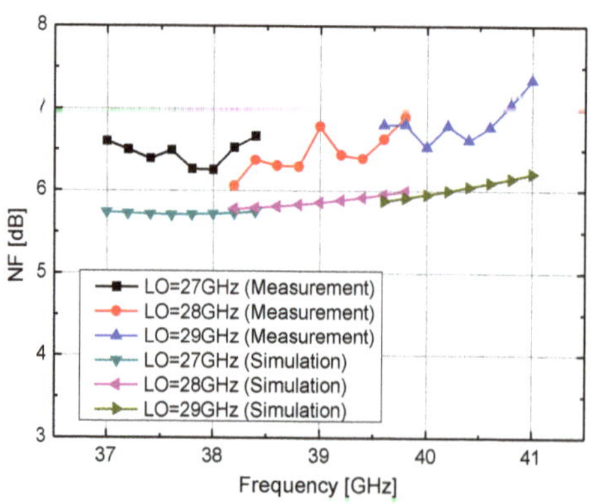

Figure 9. Measured and simulated NF of the proposed mixer. LO power = −20 dBm.

Figure 10 illustrates the simulated and measured CG of the mixer. The LO frequencies are 27 GHz, 28 GHz, and 29 GHz, while the RF frequency ranges are 37–38.25 GHz, 38.25–39.75 GHz, and 39.75–41 GHz, respectively. The RF and LO input powers are −30 dBm and −20 dBm, respectively. The measured conversion gain is 5.2–6.4 dB over an RF frequency range of 37–41 GHz. The neutralization capacitors and the separation of the Gm and switching stages contribute to enhanced gain and NF performance. Figure 11 shows the CG of the mixer for varying RF input power, with an LO input power of −20 dBm. The measured IP1dB of the mixer is −6 dBm at an RF frequency of 39 GHz.

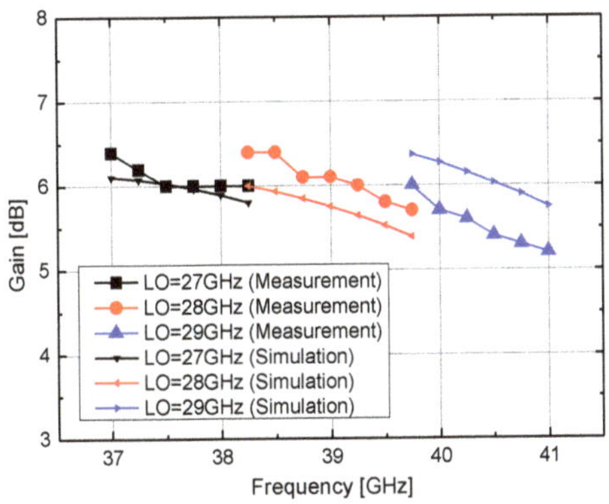

Figure 10. Measured and simulated CG of proposed mixer. RF power = −30 dBm. LO power = −20 dBm.

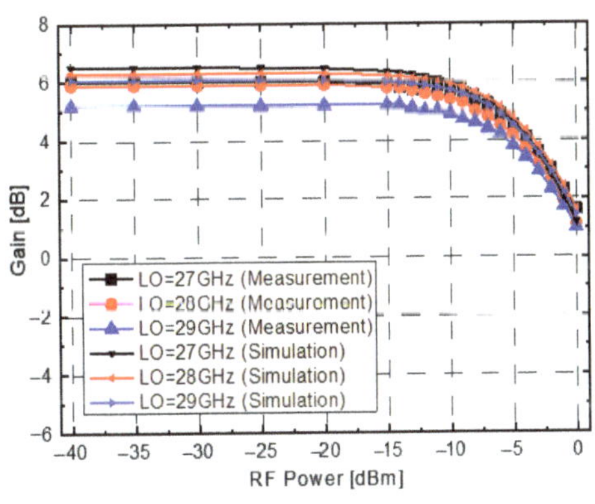

Figure 11. Measured and simulated CG versus RF input power of proposed mixer. LO power = −20 dBm. RF frequencies are 37.75/39/40.25 GHz.

Figure 12 presents the simulated and measured output third-order intercept point (OIP3) of the mixer. The measured OIP3 exceeds 9.6 dBm at an RF frequency of 39 GHz and remains above 8.9 dBm across the RF frequency range of 37–41 GHz. The measured IIP3 is greater than 3.1 dBm across the same frequency range. Simulated data show good agreement with the measurement results. Figure 13 illustrates the measured LO-to-RF and LO-to-IF isolations. LO leakage generates unwanted spurious signals by mixing with other harmonics, leading to degraded linearity and reduced signal quality. Therefore, minimizing LO leakage is crucial to maintaining optimal performance. The LO-to-RF isolation is better than 25.3 dB, while the LO-to-IF isolation exceeds 45.7 dB. Figure 14 shows the measured output spectrum. The measured harmonic suppression for an RF input power of −30 dBm is better than −26 dBc across an RF frequency range of 37–41 GHz, with the output spectrum accounting for cable loss.

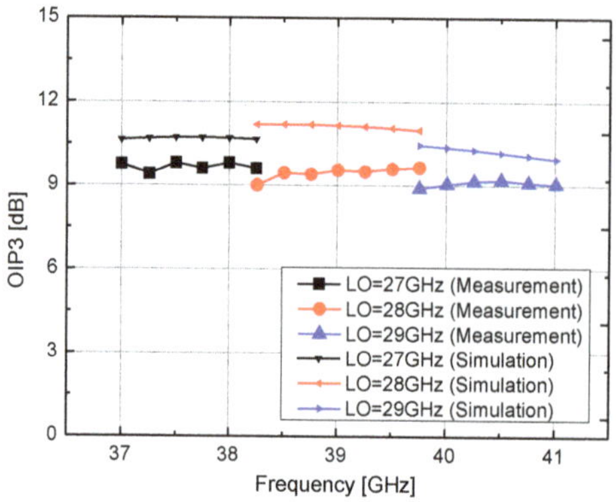

Figure 12. Measured and simulated OIP3 versus RF frequency of proposed mixer. RF power = −30 dBm. LO power = −20 dBm.

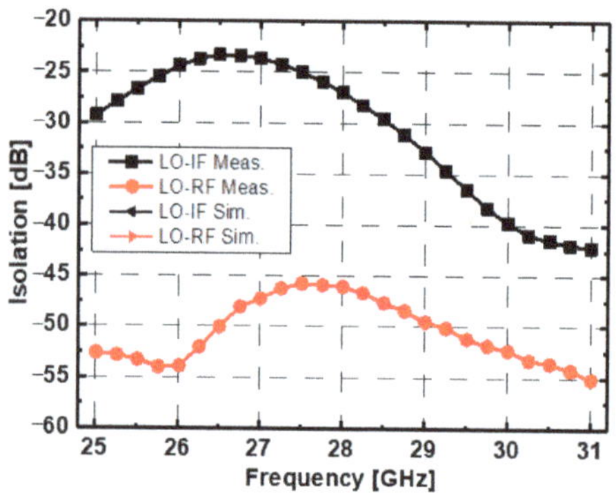

Figure 13. Measured LO-IF and LO-RF isolations of proposed mixer.

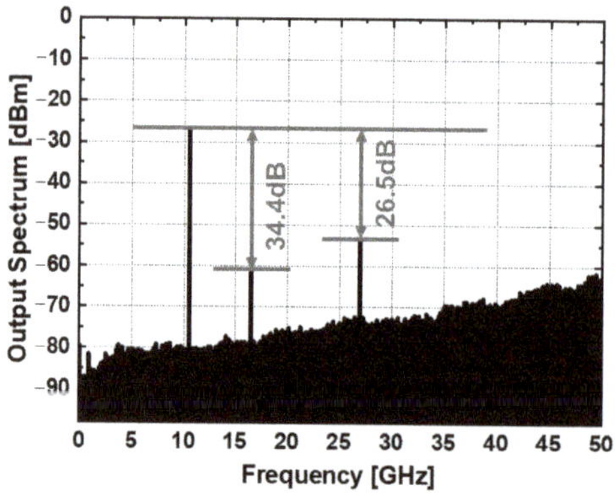

Figure 14. Measured output spectrum for harmonic suppression: f_{LO} = 27 GHz; f_{IF} = 10.5 GHz; P_{LO} = −20 dBm.

4. Discussion

Table 1 compares the performance of state-of-the-art 5G FR2 CMOS down-conversion mixers. A 5G FR2 down-conversion mixer should achieve low power consumption, high conversion gain, low noise figure, and high linearity. However, these performance parameters inherently involve trade-offs, and state-of-the-art down-conversion mixers typically prioritize two or three key aspects. The mixer utilizing MGTR at the Gm stage [22] demonstrates a high IIP3 of 20.87 dBm but suffers from a high NF of 16.75 dB and a low CG of −14.4 dB. Similarly, the mixer in [18] achieves a high CG of 12.6 dB; however, it employs a direct-conversion architecture and an IF buffer amplifier, resulting in an NF exceeding 10 dB. The proposed mixer, which incorporates neutralization capacitors, a separate Gm stage and switching stage, simultaneous gain and noise matching, and transformer-based harmonic suppression, achieves a significantly lower NF of less than 7.3 dB and a high CG of 5.2–6.4 dB, all without the need for an IF buffer amplifier. Additionally, the LO-RF isolation exceeds 46 dB, and the harmonic suppression is better than −26 dBc. The measurement results indicate that the proposed mixer is well-suited for 5G low-power mm-wave handheld devices, offering a balanced trade-off between performance and power efficiency.

Table 1. Performance comparisons of state-of-the-art.

	This Work	[15]	[16]	[17]	[18]	[20]	[21]	[22]
Process	65 nm CMOS	65 nm CMOS	40 nm CMOS	65 nm CMOS	90 nm CMOS	55 nm CMOS	22 nm CMOS	55 nm CMOS
Topology	RF Gm + Switching Core	Gilbert-cell + IF Buffer	RF Gm + Switching Core + IF Buffer	RF Gm + Switching Core + IF Buffer	Gilbert-cell + IF Buffer	Switching Core + IF Buffer	Switching Core	MGTR Gilbert-cell
Frequency (GHz)	37–41	37–40	37–40	37–40	28	19–32.8	24–34	36–42
V_{DD} (V)	1.0	1.0	1.1	1.0	0.5	1.6	0.8	1.2
CG (dB)	5.2–6.4	4.8	−4.1–1.2	3.1	12.6	−0.5	7.8	−14.4
NF (dB)	<7.3	13.5	12.4–13	11.4	10.6	-	16	16.75
LO Power (dBm)	−20	5	−10	5	0	2	−8	9
IP1dB (dBm)	−6	−5.9	−0.5	−11	-	8.2	−7	9.85
IIP3 (dBm)	3.1	0.7	4	−1.95	1.5	-	4.8	20.87
LO-RF Isolation (dB)	>46	-	-	-	37.1	45.3	>42	>32.7
P_{DC} (mW)	18	10.3	28.3	6.2	4.5	15.2	1.28	26
Area (mm^2)	0.51	0.4	0.654	0.3	0.529	0.63	0.243	1.47

5. Conclusions

A 39 GHz down-conversion mixer circuit was successfully designed and implemented using standard 65 nm CMOS technology. By incorporating neutralization capacitors, a separate Gm stage and switching stage, simultaneous gain and noise matching, and transformer-based harmonic suppression, the proposed mixer achieves a low NF of less than 7.3 dB and a high CG of 5.2–6.4 dB, all without requiring an IF buffer amplifier. Moreover, the LO-RF isolation is better than 46 dB, and the harmonic suppression exceeds −26 dBc. The measurement results validate that the proposed mixer meets the performance requirements for 5G low-power mm-wave handheld devices. Its design demonstrates the feasibility of achieving high performance with low LO input power, making it a strong candidate for next-generation wireless communication systems.

Author Contributions: Conceptualization, I.-C.Y. and C.-W.B.; data curation, I.-C.Y. and C.-W.B.; formal analysis, I.-C.Y. and C.-W.B.; funding acquisition, C.-W.B.; investigation, C.-W.B.; methodology, I.-C.Y. and C.-W.B.; project administration, C.-W.B.; resources, C.-W.B.; software I.-C.Y. and C.-W.B.; supervision, C.-W.B.; validation, I.-C.Y. and C.-W.B.; visualization, I.-C.Y. and C.-W.B.; writing—original draft, I.-C.Y.; writing—review and editing, I.-C.Y. and C.-W.B. All authors have read and agreed to the published version of the manuscript.

Funding: The EDA tool was supported by the IC Design Education Center (IDEC), Korea.

Data Availability Statement: Data are contained within this article.

Conflicts of Interest: The authors declare no conflicts of interest.

References

1. Song, J.-H.; Lee, E.-G.; Lee, J.-E.; Son, J.-T.; Kim, J.-H.; Baek, M.-S.; Kim, C.-Y. A 37–40 GHz 6-Bits Switched-Filter Phase Shifter Using 150 nm GaN HEMT. *Nanomaterials* **2023**, *13*, 2752. [CrossRef] [PubMed]
2. Kiani, S.H.; Altaf, A.; Anjum, M.R.; Afridi, S.; Arain, Z.A.; Anwar, S.; Khan, S.; Alibakhshikenari, M.; Lalbakhsh, A.; Khan, M.A.; et al. MIMO Antenna System for Modern 5G Handheld Devices with Healthcare and High Rate Delivery. *Sensors* **2021**, *21*, 7415. [CrossRef] [PubMed]
3. Hussain, S.A.; Taher, F.; Alzaidi, M.S.; Hussain, I.; Ghoniem, R.M.; Sree, M.F.A.; Lalbakhsh, A.L. Wideband, High-Gain, and Compact Four-Port MIMO Antenna for Future 5G Devices Operating over Ka-Band Spectrum. *Appl. Sci.* **2023**, *13*, 4380. [CrossRef]
4. Yin, Y.; Ustundag, B.; Kibaroglu, K.; Sayginer, M.; Rebeiz, G.M. Wideband 23.5–29.5-GHz phased arrays for multistandard 5G applications and carrier aggregation. *IEEE Trans. Microw. Theory Tech.* **2021**, *69*, 235–247. [CrossRef]
5. Yoo, I.C.; Cho, D.O.; Byeon, C.W. A Millimeter-wave CMOS Cross-Polarization Leakage Canceller for Dual-Polarized MIMO Systems. *IDEC J. Integr. Circuits Syst.* **2022**, *8*, 11–15.
6. Sadhu, B.; Paidimarri, A.; Lee, W.; Yeck, M.; Ozdag, C.; Tojo, Y. A 24-to-30GHz 256-element dual-polarized 5G phased array with fast beam-switching support for >30,000 beams. In Proceedings of the IEEE International Solid-State Circuits Conference (ISSCC), San Francisco, CA, USA, 20–26 February 2022; Volume 65, pp. 436–438.
7. Jung, J.; Lee, J.; Kang, D.; Kim, J.; Lee, W.; Oh, H.; Park, J.; Kim, K.; Lee, D.; Lee, S.; et al. A 39 GHz 2 × 16-Channel Phased-Array Transceiver IC with Compact, High-Efficiency Doherty Power Amplifiers. In Proceedings of the 2023 IEEE Radio Frequency Integrated Circuits Symposium (RFIC), San Diego, CA, USA, 11–13 June 2023; pp. 273–276.
8. Li, Z.; Pang, J.; Zhang, Y.; Yamazaki, Y.; Wang, Q.; Luo, P.; Chen, W.; Liao, Y.; Tang, M.; Wang, Y.; et al. A 39-GHz CMOS Bidirectional Doherty Phased-Array Beamformer Using Shared-LUT DPD with Inter-Element Mismatch Compensation Technique for 5G Base Station. *IEEE J. Solid-State Circuits* **2023**, *58*, 901–914. [CrossRef]
9. Park, J.; Lee, S.; Chun, J.; Jeon, L.; Hong, S. A 28-GHz Four-Channel Beamforming Front-End IC with Dual-Vector Variable Gain Phase Shifters for 64-Element Phased Array Antenna Module. *IEEE J. Solid-State Circuits* **2023**, *58*, 1142–1159. [CrossRef]
10. Wang, S.; Liang, T.; Alhamed, A.; Rebeiz, G.M. A 23–46-GHz Fully Planar 8 × 8 Multistandard 5G Phased Array With OFDM 400-MHz 64-QAM Waveforms at 40–44-dBm EIRP. *IEEE Trans. Microw. Theory Tech.* **2024**, *72*, 6739–6748. [CrossRef]
11. Lee, W.; Ozdag, C.; Plouchart, J.-O.; Valdes-Garcia, A.; Sadhu, B. A 24 to 30-GHz Phased Array Transceiver Front End With 2.8 to 3.1-dB RX NF and 22 to 24% TX Peak Efficiency. *IEEE J. Solid-State Circuits* **2024**, *59*, 2788–2804. [CrossRef]
12. Sadhu, B.; Paidimarri, A.; Watanabe, A.O.; Liu, D.; Gu, X.; Baks, C.W.; Tojo, Y.; Fujisaku, Y.; Sousa, I.D.; Yamaguichi, Y.; et al. A Heterogeneously Integrated 256-Element 5G Phased Array: Design, Assembly, Test. *IEEE J. Microwaves* **2025**, *5*, 68–83. [CrossRef]
13. Ball, E.A. Predicting the Performance of a 26 GHz Transconductance Modulated Downconversion Mixer as a Function of LO Drive and DC Bias. *Electronics* **2022**, *11*, 2516. [CrossRef]
14. Byeon, C.W.; Lee, S.H.; Lee, J.H.; Son, J.H. A high-linearity Ka-band CMOS down-conversion mixer. *Microw. Opt. Technol. Lett.* **2020**, *62*, 3785–3790. [CrossRef]
15. Bae, B.; Han, J. 24–40 GHz Gain-Boosted Wideband CMOS Down-Conversion Mixer Employing Body-Effect Control for 5G NR Applications. *IEEE Trans. Circuits Syst. II Exp. Briefs* **2022**, *69*, 1034–1038. [CrossRef]
16. Lee, D.; Lee, M.; Park, B.; Song, E.; Lee, K.; Lee, J.; Han, J.; Kwon, K. 24–40 GHz mmWave Down-Conversion Mixer With Broadband Capacitor-Tuned Coupled Resonators for 5G New Radio Cellular Applications. *IEEE Access* **2022**, *10*, 16782–16792. [CrossRef]
17. Kim, E.; Kim, S.; Kim, G.; Han, J. Dual-Band CMOS Down-Conversion Mixer Adopting Band-Switchable Transformer. *IEEE Trans. Circuits Syst. II Exp. Briefs* **2023**, *70*, 3902–3906. [CrossRef]
18. Lin, Y.-S.; Lan, K. S. Down-Conversion Mixer Using λ/4-TL-C-based Coupler and BSFB Technique for 28 GHz 5G NR. In Proceedings of the 2023 IEEE/MTT-S International Microwave Symposium (IMS), San Diego, CA, USA, 11–16 June 2023; pp. 1128–1131.
19. Fu, S.; Zhang, X.; Liu, B.; Shi, C.; Huang, L.; Chen, J.; Zhang, R. A High-Gain and Low-Noise Mixer with Hybrid Gm-Boosting for 5G FR2 Applications. In Proceedings of the 2023 IEEE International Symposium on Circuits and Systems (ISCAS), Monterey, CA, USA, 21–25 May 2023; pp. 1–5.
20. Chen, Z.; Tian, W.; Huang, S.; Zhang, X.; Zhang, Y.; Li, X.; Lu, M.; Hu, J.; Ouyang, K.; Long, Z. A 19-32.8 GHz Low Power Down-Conversion Mixer with 8.2 dBm IP1dB for 5G Communication. In Proceedings of the 2024 54th European Microwave Conference (EuMC), Paris, France, 24–26 September 2024; pp. 51–54.
21. Dossanov, A.; Issakov, V. A 1.28mW K-Band Modified Gilbert-Cell Mixer Design in 22nm FDSOI CMOS. In Proceedings of the 2024 IEEE Topical Meeting on Silicon Monolithic Integrated Circuits in RF Systems (SiRF), San Antonio, TX, USA, 21–24 January 2024; pp. 110–112.

22. Hu, K.; Ma, K.; Ma, Z.; Wang, K. A Reusable Superheterodyne Dual-Band Down-Conversion Mixer With Hybrid Linearity-Enhanced Technique for 5G Non-Contiguous Multiband NR. *IEEE J. Solid-State Circuits* **2024**, *59*, 3392–3405. [CrossRef]
23. Razavi, B. Design of millimeter-wave CMOS radios: A tutorial. *IEEE Trans. Circuits Syst. I* **2009**, *56*, 4–16. [CrossRef]
24. Byeon, C.W.; Lee, J.H.; Lee, S.H.; Son, J.H. A Ka-band variable-gain amplifier with low OP1dB variation for 5G applications. *IEEE Microw. Wireless Compon. Lett.* **2019**, *29*, 722–724. [CrossRef]
25. Kim, J.-H.; Byeon, C.-W. A 60 GHz Power Amplifier with Neutralization Capacitors and Compensation Inductors. *Electronics* **2024**, *13*, 4276. [CrossRef]
26. Long, J.R. Monolithic transformers for silicon RF IC design. *IEEE J. Solid-State Circuits* **2000**, *35*, 1368–1382. [CrossRef]
27. Kuo, C.; Tsai, Z.; Tsai, J.; Wang, H. A 71–76 GHz CMOS variable gain amplifier using current steering technique. In Proceedings of the 2023 IEEE Radio Frequency Integrated Circuits Symposium (RFIC), Atlanta, GA, USA, 15–17 June 2008; pp. 609–612.

electronics

MDPI

Article

A 60 GHz Power Amplifier with Neutralization Capacitors and Compensation Inductors

Joon-Hyung Kim [1] and Chul-Woo Byeon [2,*]

[1] Department of Semiconductor Convergence, Chungnam National University, Daejeon 34134, Republic of Korea; joonhyung.kim@cnu.ac.kr

[2] School of Electronics and Electrical Engineering, College of Engineering, Dankook University, Yongin-si 16890, Republic of Korea

* Correspondence: cwbyeon@dankook.ac.kr; Tel.: +82-31-8005-3632

Abstract: In this paper, we present a high power-added efficiency (PAE) and high gain per stage 60 GHz power amplifier (PA). The proposed PA consists of a two-stage common-source amplifier that incorporates neutralization capacitors and compensation inductors to enhance both gain and efficiency. The gain characteristics are analyzed, demonstrating that the proposed design improves both gain and efficiency. Implemented in 65 nm CMOS technology, the PA achieves a saturated output power of 13.4 dBm at 60 GHz, with a maximum PAE of 26.7% from a 1 V supply. The output 1 dB compression point is 10.5 dBm, with a PAE of 16%. The PA occupies a core chip area of 0.094 mm^2.

Keywords: 60 GHz; CMOS; gain; millimeter wave; neutralization capacitor; power-added efficiency; power amplifier; transformer

1. Introduction

Recently, the unlicensed 60 GHz band has gained attention for its ability to transmit tens of gigabits per second of data due to its wide bandwidth [1–3]. In this frequency band, low-power, high-efficiency designs are critical for handheld device applications. Additionally, multiple-input multiple-output (MIMO) and phased array techniques have been introduced to enhance data rates and extend communication distances. However, these techniques require multiple transmitters and receivers, leading to increased power consumption, which makes their use in handheld devices challenging. In particular, power amplifiers (PAs) consume substantial DC power while offering limited efficiency.

As a key component, PAs with high power-added efficiency (PAE) and output power exceeding 13 dBm are essential for reducing battery consumption and extending the operating time of handheld devices. Thus, designing highly efficient PAs is a primary challenge for next-generation millimeter-wave (mm-wave) wireless systems. PA efficiency represents a significant bottleneck, especially in transmitter arrays. Over the past decade, significant progress has been made in developing high-power, high-efficiency PAs for mm-wave bands [4–14]. CMOS processes are widely used in mm-wave PA design due to their low cost and ability to integrate with digital baseband circuitry. Transformer-based common-source amplifiers are popular for their compact size and simple power matching and combining capabilities [7–14]. Most of the published CMOS PAs focused on achieving high output power by employing cascode topologies [8,9], power combining architectures [10–12], and linearity enhancement techniques [13]. Power combining techniques enable higher output power and gain [10–12], but this technique suffers from large chip areas and limited PAE. For instance, a PA with a two-way, four-stage common-source [10] achieved a saturated output power (PSAT) of 16 dBm, but with a PAE of only 17%. A 32-way, four-stage common-source PA [11] showed a gain of 21.9 dB and a PSAT of 24.1 dBm but required a large chip area and a huge power consumption. The linearized PA [13] achieved a PSAT of 15 dBm with a PAE of 26%, and an output 1 dB compression point (OP1dB) of 14 dBm with a PAE of

21%. However, this PA exhibited a limited gain of 14 dB due to the linearization technique. A variable gain PA [14] provides a variable gain of 23.5 dB, but requires multiple stages, limiting its PAE to 21.5%.

In this study, a high-efficiency 60 GHz PA was implemented for handheld devices such as smart phones, gaming consoles, augmented reality, and virtual reality devices. The proposed circuit uses a two-stage common-source amplifier with compensation inductors and neutralization capacitors to achieve high PAE and high gain per stage. As a result, the PA successfully achieved a power gain of 18.3 dB, an OP1dB of 10.5 dBm, and a peak PAE (PAE_{peak}) of 26.7% at 60 GHz. The measurement results indicate that the proposed PA is suitable for low-power mm-wave handheld devices.

2. Design Methodology

2.1. Active Device

The output power and efficiency of the PAs highly depend on active devices and parasitic components, as PAs typically utilize large transistors to achieve high output power. This issue becomes more pronounced in the mm-wave band, where operating frequencies approach the cut-off frequency and maximum oscillation frequency, leading to reductions in power gain and output power. Additionally, the parasitic capacitance of active devices further limits bandwidth, power gain, output power, and PAE. PAE can be expressed as

$$PAE = \frac{P_{OUT} - P_{IN}}{P_{DC}}, \tag{1}$$

where P_{OUT}, P_{IN}, and P_{DC} are the output power, input power, and power dissipation of the PA. PAE is a metric for rating the efficiency of the PA. To mitigate these challenges, careful design of transistor size and interconnections is essential.

To optimize the transistor size for the PA, we compared the performance of the power stage based on different device configurations. Table 1 presents the load-pull simulation results comparing transistors with one 4 μm × 25 fingers, two 2 μm × 25 fingers, and four 1 μm × 25 fingers in a differential common-source amplifier with neutralization capacitors [7]. All configurations maintain a total transistor width of 100 μm to achieve an output power above 13 dBm. The simulation results indicate that the configuration with two 2 μm × 25 fingers delivers superior PSAT and PAE_{peak}. As a result, we selected two 2 μm × 25 fingers for the power cell in the power stage of this design.

Table 1. Performance comparison of load-pull simulation results at 60 GHz.

Topology	4 μm × 25 Fingers	2 × 2 μm × 25 Fingers	4 × 1 μm × 25 Fingers
PSAT (dBm)	15.2	15.9	15.5
PAE_{peak} (%)	33.8	37.5	36.5
Power Gain (dB)	11.1	12.0	12.2
Z_{opt} (Ω)	21.1 + j15.3	16.3 + j16.5	17.4 + 21.0

Figure 1a illustrates the proposed layout of the unit transistor cell for the drive and power stages. The unit cell employs transistors with a width of 2 μm and 25 fingers, occupying an area of 15 × 9 μm^2. To optimize the layout, the source is connected using multiple metal layers (M4–M7) and linked to a double-sided ground plane (M1–M2). The M1 and gate poly are connected on both sides and routed to M8. Additionally, gate–source and gate–drain overlap is minimized to reduce parasitic capacitance. Figure 1b shows the layout of the two 2 μm × 25 fingers devices, which occupy an area of 28 × 9 μm^2. The unit transistors are placed adjacent to each other, with the sources directly connected to minimize interconnection resistance. The gate and drain are routed through thick metal layers M8 and M9, respectively. In electromagnetic (EM) simulations, interconnection losses resulted in power gain degradations of 0.2 dB and 0.3 dB for the single 2 μm × 25 fingers and two 2 μm × 25 fingers transistors, respectively.

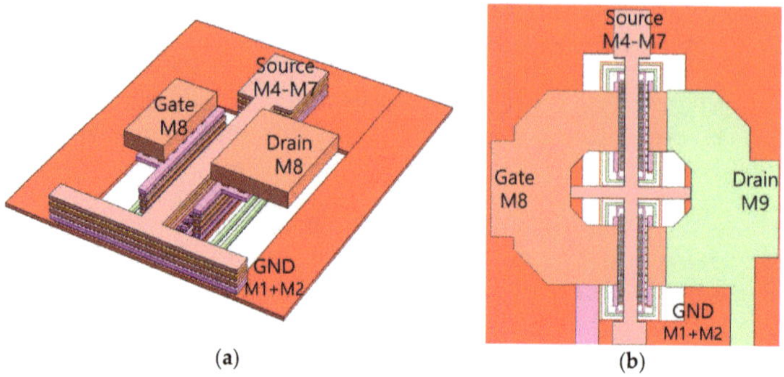

Figure 1. Transistor layout of (**a**) 2 μm × 25 fingers and (**b**) 2 × 2 μm × 25 fingers.

2.2. Neutralization Capacitors

In addition to optimizing transistor configuration and layout, neutralization capacitors [7] are employed to enhance gain and stability. Figure 2 shows the schematic and layout of the differential amplifier with neutralization capacitors for the power stage. The gate-to-drain parasitic capacitance, C_{gd}, creates a feedback path that degrades power gain, reverse isolation, and stability. The neutralization capacitors, C_N, address these issues by neutralizing C_{gd}, effectively eliminating the feedback path.

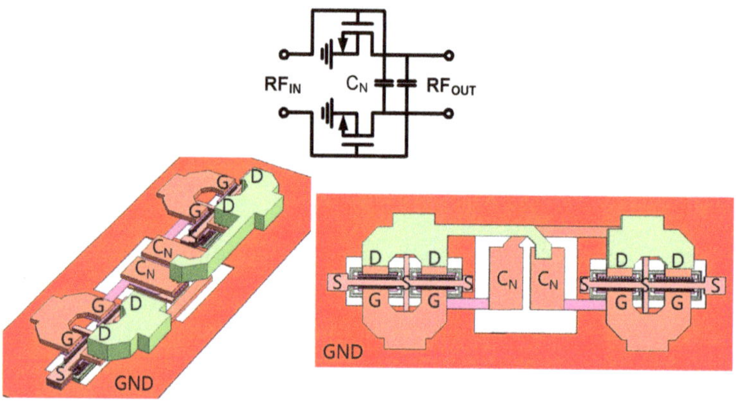

Figure 2. Schematic and layout of differential amplifier with neutralization capacitors, where G, D, and S are gate, drain, and source, respectively.

The custom-designed metal–oxide–metal (MOM) capacitor [15] was used for C_N, rather than the standard foundry-provided MOM capacitor. The custom MOM capacitor utilizes a multi-stack metal plate design for a smaller chip area, shorter interconnections, and sufficient vias to reduce parasitic resistance. Moreover, the custom MOM capacitor exhibited a minimal process variation of less than ±2%, compared to ±11% for the foundry-provided MOM capacitor, as shown in simulations.

Figure 3 illustrates the simulated maximum available gain (MAG) and stability factor of the differential common-source amplifier with the neutralization capacitor C_N. The amplifier remains stable with up to ±20% variation in C_N. A capacitance of 27 fF was selected for C_N, slightly smaller than the required value for maximum stability, to account for additional parasitic capacitance. With the neutralization technique, the power stage achieves a MAG of 13.1 dB, an improvement of 3.0 dB compared to the design without neutralization. The neutralization capacitor is also applied to the driver stage to further enhance stability and gain.

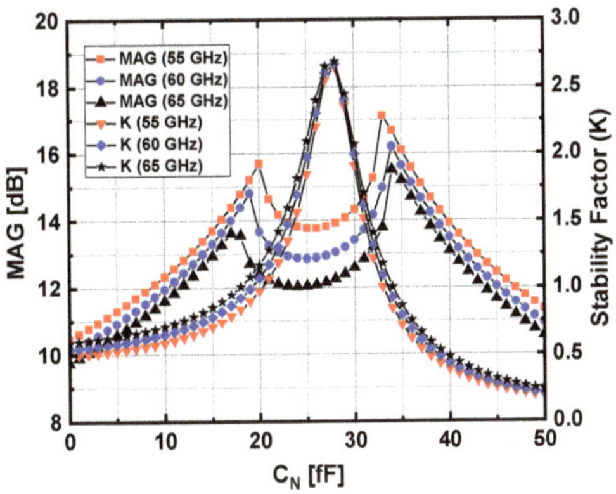

Figure 3. Simulated MAG and stability factor of the differential common-source amplifier with neutralization capacitor C_N.

2.3. *Proposed PA Design with Transformers and Compensation Inductors*

A simplified schematic of the proposed 60 GHz PA and the layout of the transformers are shown in Figure 4. The PA consists of a two-stage differential common-source amplifier with transformers, TF_1–TF_3, a matching network with compensation inductors, and neutralization capacitors, C_{N1} and C_{N2}. The transistor size of the power stage is two times larger than that of the driver stage. The power stage has two 2 μm × 25 fingers devices, whereas the driver stage has a single 2 μm × 25 fingers device. In the power stage, two 2 μm × 25 fingers devices provide a better gain, PAE, and PSAT than a single 4 μm × 25 fingers device. The neutralization capacitors C_{N1} and C_{N2} are chosen to be 13.5 fF and 27 fF, respectively, for better power gain, isolation, and stability.

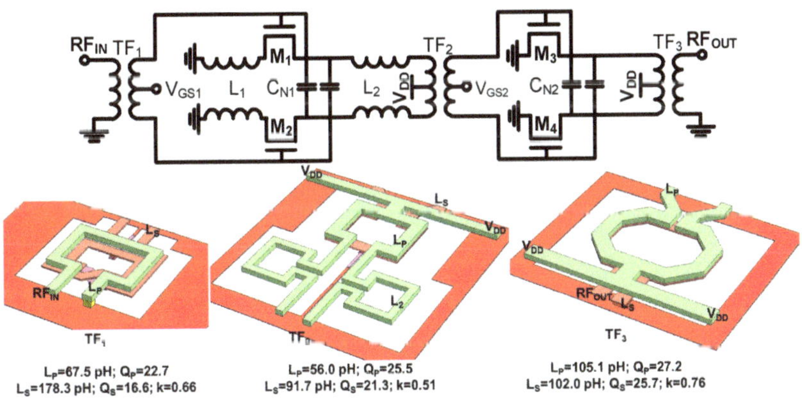

Figure 4. Schematic of proposed PA and layout of transformers.

The first stage of the PA is implemented using a differential common-source amplifier with neutralization capacitors and inductive source degeneration. Inductor L_1 is connected to the source terminals of the transistors M1 and M2, creating a series–series negative feedback loop. This feedback improves OP1dB and stability while slightly reducing power gain. The inductive source degeneration technique at the power stage also enhances the linearity, but the gain reduction at the second stage requires a higher linearity at the first stage. L_1 is set to 24 pH to balance the improvement in OP1dB and stability with the reduction in gain. In the simulation, a 0.5 dB improvement in OP1dB is achieved, with a 1.5 dB reduction in gain, while the stability factor increases from 10 to 15 at 60 GHz.

A highly efficient matching network is crucial for enhancing gain, PAE, and PSAT. Transformer-based matching networks, paired with compensation inductors, are used to achieve high efficiency and a compact design. The transformers are implemented using two vertically coupled top metal layers. The input transformer balun TF_1 has a turn ratio of 1:2 for impedance transformation. The output transformer TF_3 is designed with a high coupling factor of 0.76, as the insertion loss of the transformer is proportional to the square of the coupling factor [16]. The inter-stage matching network uses compensation inductors L_2 and a transformer, TF_2, with a lower coupling coefficient and a 1:1 turn ratio. A coupling coefficient of 0.51 was selected for TF_2 considering the matching loss and bandwidth. A lower coupling coefficient results in wider bandwidth but increases matching loss [4]. The compensation inductor L_2 helps optimize the matching loss. As shown in Figure 5, the simulated inter-stage matching loss is 1.9 dB at 60 GHz. The inter-stage matching network, which combines the moderately coupled transformer TF_2 with the compensation inductor L_2, delivers high efficiency and gain by compensating the impedance mismatch between the driver-stage output and the power-stage input. Consequently, the matching loss of the inter-stage matching network with TF_2 and L_2 is 3.4 dB smaller than that with the transformer without a compensation inductor in the simulation.

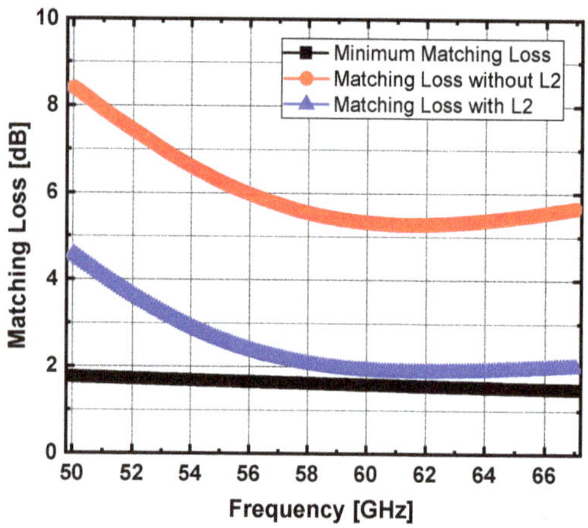

Figure 5. Simulated inter-stage matching losses.

A ground plane with a mesh pattern using the two bottom metal layers is used. The neutralization capacitors, inductors, interconnections, and transformers are all simulated using the HFSS 3-D EM simulator developed by Ansys, Canonsburg, PA, USA. The transistors, M1–M4, have a gate length of 60 nm. With the neutralization capacitors and compensation inductors, the proposed PA achieved a simulated gain of 17.6 dB, a simulated PSAT of 14 dBm, and a simulated PAE_{peak} of 26.5% at 60 GHz.

3. Results

The proposed PA circuit was implemented in a standard 65 nm CMOS technology. Figure 6 depicts a chip photograph of the proposed PA. The PA chip occupies an area of 0.58×0.46 mm^2 including the bond pads and 0.094 mm^2 excluding the pads. The PA dissipates a power of 58 mW from a 1 V supply voltage. Figure 7 illustrates measurement setups for the S-parameter and power handling capability. The S-parameter was measured with an E8361A vector network analyzer manufactured by Keysight, Santa Rosa, CA, USA and the results were calibrated using the Picoprobe SOLT (short–open–load–thru) GSG-67A calibration substrate. The large-signal measurements were performed with a signal generator, frequency multiplier, power sensor, and high-precision power meter. The measured losses of the probe tips, adapters, and coaxial cables were de-embedded from the raw measurement data.

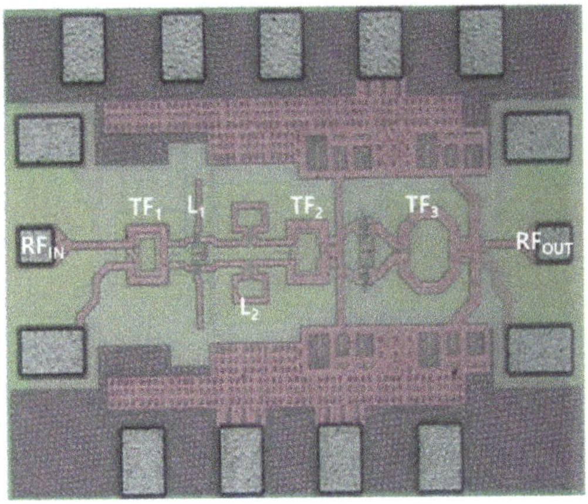

Figure 6. Microphotograph of proposed PA.

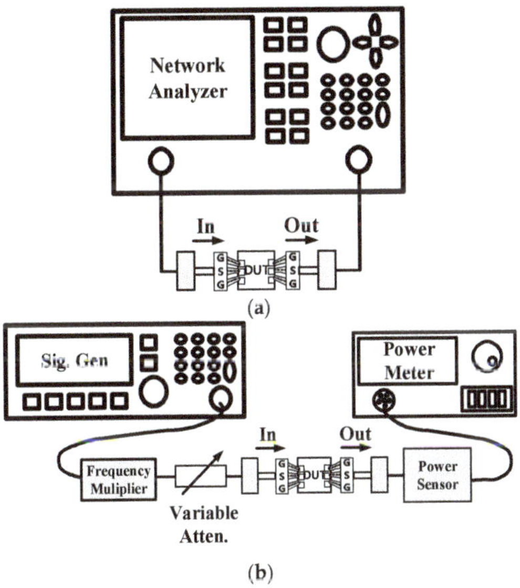

Figure 7. Measurement setups for (**a**) S-parameter and (**b**) large-signal measurements.

Figure 8 shows the simulated and measured power gains of the 60 GHz PA. The measured gain of the proposed PA is 18.3 dB at 60 GHz. The measured peak gain is 18.7 dB at 58.3 GHz and the measured 3 dB bandwidth is 10.1 GHz (54.4–64.5 GHz). The difference between the simulated and measured data is probably due to the inaccuracy of the transistor model at 60 GHz, especially the drain-to-gate capacitance. Figure 9 depicts the simulated and measured input and output return losses of the proposed PA. The measured input and output return losses are larger than 8.4 dB and 6.0 dB, respectively, for 55.2–66 GHz. Figure 10 illustrates the simulated and measured reverse isolation of the proposed PA. The reverse isolation is smaller than −39 dB in the 60 GHz band.

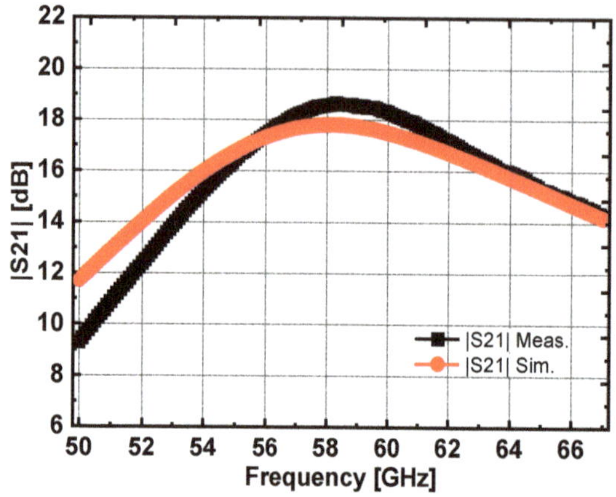

Figure 8. Simulated and measured power gains of the proposed PA.

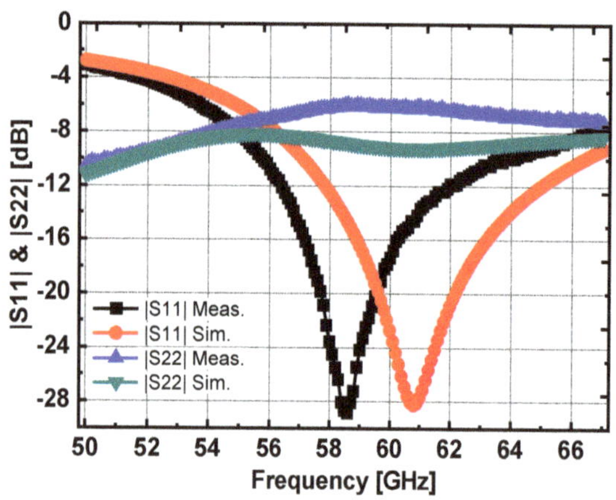

Figure 9. Simulated and measured input and output return losses of the proposed PA.

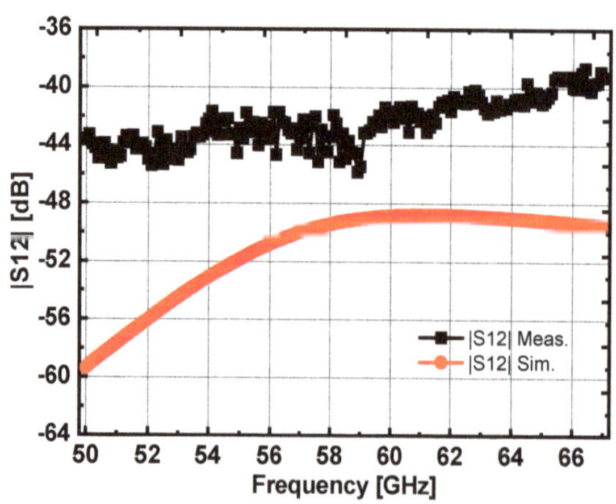

Figure 10. Simulated and measured reverse isolation of the proposed PA.

Figure 11 shows the measured power gain, output power, and PAE. OP1dB and PSAT are 10.5 dBm and 13.4 dBm at 60 GHz, respectively. The maximum PAE is 26.7%. The measured PSAT is limited due to the limited input power of the signal generator and frequency multiplier. The neutralization capacitors and compensation inductors enhance the gain and PAE, which are suitable for low-power mm-wave handheld devices.

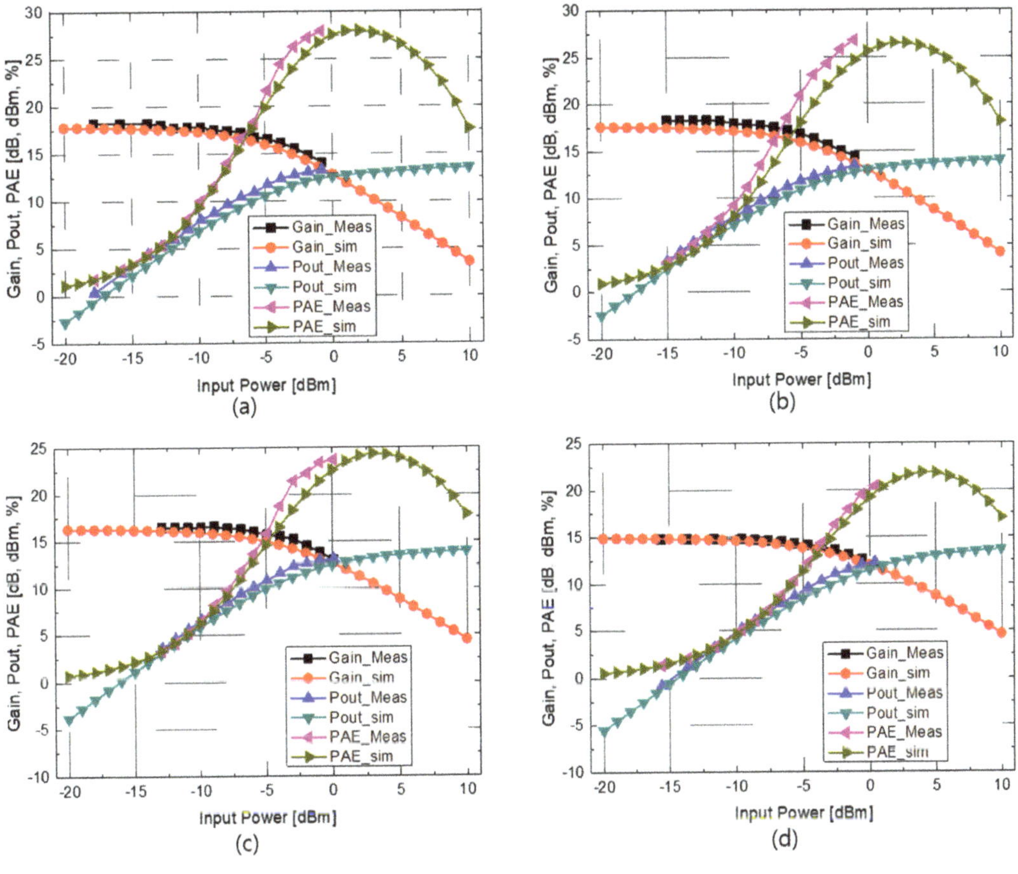

Figure 11. Simulated and measured power performances of the proposed PA at (**a**) 57 GHz, (**b**) 60 GHz, (**c**) 63 GHz, and (**d**) 66 GHz.

4. Discussion

Table 2 compares the performance of the state-of-the-art 60 GHz CMOS PAs. The proposed PA demonstrates high gain per stage and a high peak power-added efficiency (PAE). The PA in [8] achieves a high PAE, but it is implemented using 22 nm CMOS technology, which offers a significantly higher cut-off frequency and maximum oscillation frequency. The PA in [11] attains high output power, due to the use of larger transistors; however, this results in a substantial DC power consumption of 942 mW. The proposed PA utilized the neutralization capacitors and compensation inductors and enhanced the gain and PAE, which are suitable for low-power mm-wave handheld devices.

Table 2. Performance comparisons of the state-of-the-art 60 GHz CMOS PA.

	This Work	**[8]**	**[9]**	**[10]**	**[11]**	**[12]**	**[13]**	**[14]**
Process	65 nm CMOS	22 nm CMOS	65 nm CMOS	65 nm CMOS	65 nm CMOS	16 nm CMOS	40 nm CMOS	55 nm CMOS
Topology	1-way 2-stage CS	1-way 2-stage cascode	1-way 4-stage CS cascode	2-way 4-stage CS	32-way 4-stage CS	4-way 2-stage CS	1-way 4-stage CS	1-way 4-stage CS
Frequency (GHz)	60	62	60	60	60	65	60	60
V_{DD} (V)	1.0	2	1.2	1.2	1.2	0.95	2	1.2
Gain (dB)	18.3	23.3	25.4	25	21.9	18 *	14	22.5
Gain/stage (dB)	9.2	11.7	6.4	6.3	5.4	9.0	7.0	5.6
PSAT (dBm)	13.4	13.6	12.2	16	25.1	17.9	15	14.4
P1dB (dBm)	10.5	12.1	11.3	11.3	21.7	13.5	14	13
PAE_{peak} (%)	26.7	28.3	8.9	17	21	26.5	26	21.5
PAE_{1dB} (%)	16	26 *	6.5 *	7 *	7 *	15	21 *	17.7
P_{DC} (mW)	58	33	186	185	942 *	-	110	67.2
Area (mm^2)	0.094	0.065	0.36	0.20	1.17	0.107	0.2	0.15

* Graphically estimated.

5. Conclusions

A 60 GHz PA circuit was implemented and designed using a standard 65 nm CMOS technology. The PA with neutralization capacitors and compensation inductors provides a gain of 18.3 dB with a maximum PAE of 26.7% at 60 GHz from a 1 V supply voltage. Furthermore, the PSAT is 13.4 dBm. The measurement results indicate that the proposed PA is suitable for application to low-power mm-wave handheld devices.

Author Contributions: Conceptualization, J.-H.K. and C.-W.B.; data curation, J.-H.K. and C.-W.B.; formal analysis, J.-H.K. and C.-W.B.; funding acquisition, C.-W.B.; investigation, C.-W.B.; methodology, J.-H.K. and C.-W.B.; project administration, C.-W.B.; resources, C.-W.B.; software J.-H.K. and C.-W.B.; supervision, C.-W.B.; validation, J.-H.K. and C.-W.B.; visualization, J.-H.K. and C.-W.B.; writing—original draft, J.-H.K.; writing—review and editing, J.-H.K. and C.-W.B. All authors have read and agreed to the published version of the manuscript.

Funding: The present research was supported by the research fund of Dankook University in 2023.

Data Availability Statement: Data are contained within this article.

Conflicts of Interest: The authors declare no conflicts of interest.

References

1. Wang, Y.; Liu, B.; Wu, R.; Liu, H.; Narayanan, A.T. A 60-GHz 3.0-Gb/s Spectrum Efficient BPOOK Transceiver for Low-Power Short-Range Wireless in 65 nm CMOS. *IEEE J. Solid-State Circuits* **2019**, *54*, 1363–1374. [CrossRef]
2. Byeon, C.W.; Eun, K.C.; Park, C.S. A 2.65-pJ/Bit 12.5-Gb/s 60-GHz OOK CMOS Transmitter and Receiver for Proximity Communications. *IEEE Trans. Microw. Theory Tech.* **2020**, *68*, 2902–2910. [CrossRef]
3. Wang, Y.; Chung, H.; Ma, Q.; Rebeiz, G.M. A 57.5–65.5 GHz Phased-Array Transmit Beamformer in 45 nm CMOS SOI with 5 dBm and 6.1% Linear PAE for 400 MBaud 64-QAM Waveforms. *IEEE Trans. Microw. Theory Tech.* **2021**, *69*, 1772–1779. [CrossRef]
4. Son, H.S.; Jang, T.H.; Kim, S.H.; Jung, K.P.; Kim, J.H.; Park, C.S. Pole-controlled wideband 120 GHz CMOS power amplifier for wireless chip-to-chip communication in 40-nm CMOS process. *IEEE Trans. Circuits Syst. II Exp. Briefs* **2019**, *66*, 1351–1355. [CrossRef]
5. Kim, J. Broadband Millimeter-Wave Power Amplifier Using Modified 2D Distributed Power Combining. *Electronics* **2020**, *9*, 899. [CrossRef]
6. Mayeda, J.; Lie, D.Y.C.; Lopez, J. Broadband Millimeter-Wave 5G Power Amplifier Design in 22 nm CMOS FD-SOI and 40 nm GaN HEMT. *Electronics* **2022**, *11*, 683. [CrossRef]

7. Chan, W.L.; Long, J.R. A 58–65 GHz neutralized CMOS power amplifier with PAE above 10% at 1-V supply. *IEEE J. Solid-State Circuits* **2010**, *45*, 554–564. [CrossRef]
8. Xu, X.; Wagner, J.; Ellinger, F. A 62-GHz High-Efficiency Power Amplifier with Modulation Capability via Back-Gate in 22-nm FD-SOI. *IEEE Solid-State Circuits Lett.* **2023**, *6*, 49–52. [CrossRef]
9. Feng, J.; Lu, L.; Chen, X.; Sun, Y.; Wu, X.; Fan, X.; Li, L. A 60-GHz Variable-Gain Power Amplifier with a Switchable Inductor for TDD Systems. *IEEE Microw. Wireless Tech. Lett.* **2024**, *34*, 512–515. [CrossRef]
10. Feng, J.; Liang, Y.; Niu, X.; Lu, L.; Chen, X.; Cheng, D.; Chen, Q.; Luo, L.; Wu, X.; Fan, X.; et al. A Compact Low-Loss High-Reliability Antenna T/R Switch Embedded in Power Combiner for 60-GHz Fully Differential PA and LNA. *IEEE Microw. Wireless Tech. Lett.* 2024; *to be appeared*.
11. Zhang, L.; Ma, K.; Fu, H. A 60-GHz 32-Way Hybrid Power Combination Power Amplifier in 55-nm Bulk CMOS. *IEEE Trans. Microw. Theory Tech.* **2023**, *71*, 613–627. [CrossRef]
12. Chu, K.-D.; Callender, S.; Wang, Y.; Rudell, J.C.; Pellerano, S.; Hull, C. A dual-mode V-band 2/4-way non-uniform power-combining PA with +17.9-dBm PSAT and 26.5-% PAE in 16-nm FinFET CMOS. In Proceedings of the 2020 IEEE Radio Frequency Integrated Circuits Symposium (RFIC), Los Angeles, CA, USA, 4–6 August 2020; pp. 183–186.
13. Mosalam, H.; Xiao, W.; Gui, X.; Li, D.; Pan, Q. A 54–68 GHz Power Amplifier with Improved Linearity and Efficiency in 40 nm CMOS. *IEEE Trans. Circuits Syst. II Exp. Briefs* **2022**, *69*, 40–44. [CrossRef]
14. Li, X.; Cheng, D.; Jiang, X.; Wang, D.; Li, L. A 57–71-GHz Accurate dB-Linear Variable Gain Power Amplifier with Ultralow Gain Error Using Particle Swarm Optimization Algorithm. In Proceedings of the 2024 IEEE/MTT-S International Microwave Symposium, Washington, DC, USA, 16–21 June 2024; pp. 454–457.
15. Byeon, C.W.; Park, C.S. A low-loss compact 60-GHz phase shifter in 65-nm CMOS. *IEEE Microw. Wireless Compon. Lett.* **2017**, *27*, 663–665. [CrossRef]
16. Aoki, I.; Kee, S.D.; Rutledge, D.B.; Hajimiri, A. Distributed active transformer—A new power-combining and impedance-transformation technique. *IEEE Trans. Microw. Theory Tech.* **2002**, *50*, 316–331. [CrossRef]

electronics

MDPI

Article

Interstacked Transformer Quad-Core VCOs

Daniele Tripoli [1], Giorgio Maiellaro [1], Santi Concetto Pavone [2] and Egidio Ragonese [2,*]

[1] STMicroelectronics, 95121 Catania, Italy; daniele.tripoli@st.com (D.T.); giorgio.maiellaro@st.com (G.M.)
[2] Dipartimento di Ingegneria Elettrica Elettronica e Informatica (DIEEI), University of Catania, 95125 Catania, Italy; santi.pavone@unict.it
* Correspondence: egidio.ragonese@unict.it; Tel.: +39-095-738-2331

Abstract: This paper presents for the first time a quad-core oscillator based on a very compact interstacked transformer that tightly couples the four cores without oscillation mode ambiguity thanks to its strong magnetic coupling factor. As a proof of concept, a 19.125 GHz oscillator for a narrowband 77 GHz radar system was designed in 28 nm fully depleted silicon-on-insulator CMOS technology with a general purpose back-end-of-line. The soundness of the proposed quad-core oscillator topology is demonstrated by comparison with state-of-the-art quad-core solutions, highlighting a significant advantage in terms of area occupation and power consumption. The proposed topology can be profitably exploited in several RF/mm-wave applications, such as radar and wireless communication systems.

Keywords: CMOS integrated circuits; fully depleted silicon on insulator; interleaved transformers; interstacked transformers; phase noise; power consumption; quad-core oscillator; stacked transformers

1. Introduction

Radio frequency (RF) and mm-wave applications, such as radar (radio detecting and ranging) and 5G, require the generation of frequency-modulated signals that are typically realized by a voltage-controlled oscillator (VCO), driven by a modulator, inside a phaselocked-loop (PLL). The VCO is the key block, especially in applications needing stringent phase noise performance. For instance, accuracy and resolution of a radar sensor are strictly related to the phase noise, thus a low-phase-noise oscillator is mandatory to achieve the highest possible target discrimination [1,2]. In this context, LC-resonant VCOs are highly preferred at the cost of a large silicon area consumption, mainly due to the tank inductor [3–8]. Traditionally, RF/mm-wave ICs are implemented in BiCMOS or CMOS technologies [9–14]. Despite several advantages of BiCMOS over CMOS in terms of noise (i.e., lower flicker noise corners), thicker back-end-of-line (BEOL) [15], higher transistor breakdown voltage (BV), and lower transconductance, g_m, at a given current level, CMOS is becoming the reference process since it is highly suitable for system-on-chip (SoC) integration, which is pursued by microelectronic industries to reduce chip cost, power consumption, and area occupation [16,17]. Unfortunately, the transition from BiCMOS to CMOS requires proper topologies and design approaches for almost all main RF/mm-wave front end blocks. In the last few years, several techniques have been developed to minimize phase noise and improve overall performance of CMOS VCOs. Specifically, multi-core techniques, which consist of coupling multiple in-phase oscillators, are theoretically powerful solutions for CMOS low-phase-noise VCO design. Indeed, the phase noise of multi-core oscillators theoretically decreases by $10 \times \log(N)$, where N is the number of cores [18–21]. However, existing implementations achieved significant phase noise reduction at the expense of power consumption and silicon area. These two parameters are often the limit to the adoption of a solution in a commercial product. Current state-of-the-art solutions to mitigate this problem involve the use of stacked inductors to couple the cores [22]. While this allows reducing the area occupation, it is still not an optimal solution for quad-core oscillators, due to the intrinsic asymmetry between the cores and the need to have two passive structures.

This work presents a novel solution based on an interstacked transformer topology, which is highly suitable for a quad-core implementation of a VCO and allows significantly reducing both area occupation and power consumption. The paper is organized as follows. A brief review of the multi-core technique is provided in Section 2 to highlight benefits and drawbacks of state-of-the-art solutions, while Section 3 introduces the interstacked topology in comparison with traditional configurations. Sections 4 and 5 are focused on the design of an interstacked transformer and the quad-core oscillator, respectively. Finally, conclusions are drawn in Section 6.

2. Multi-Core Oscillator Review

The best design strategy for phase noise minimization of LC-resonant VCOs is to use a low inductance value, L, while maintaining a sufficiently high tank quality factor, Q_T (i.e., minimize the L/Q_T ratio). However, low inductance values are related to higher losses in the tank, which means that optimizing the phase noise performance could be in contrast with the power consumption. Therefore, the L/Q_T ratio minimization must be pursued by a tradeoff between the inductance value, L, and the resulting tank quality factor, Q_T. Unfortunately, two issues hinder the above design strategy in practical implementations. First, low inductance coils are highly sensitive to layout parasitics. Second, inductance decrease would imply a consequent increment of the overall required capacitance, which would intrinsically exhibit lower-quality factor, Q_C, especially at mm-wave frequencies, with a resulting Q_T degradation. Given the integration technology, these constraints set a limit to the minimum phase noise that a single oscillator can achieve.

The multi-core technique allows overcoming the above-described limitations and pursuing L/Q_T ratio minimization. The main idea is to couple multiple oscillators (i.e., cores) together by connecting them through a generic impedance network to put their output voltages in phase. The coupling network can be resistive, capacitive, or inductive. In general, a complex impedance network can be realized. A theoretical analysis of different types of coupling and their impact on oscillator performance is available in [20]. Ideally, at steady state, no current flows through the coupling network, and all outputs are virtually "shorted". To achieve a robust coupling design, the coupling impedance must be low-Q, and as low as possible to guarantee enough suppression of undesired oscillation modes and a strong coupling between the cores, respectively.

It can be easily demonstrated that the phase noise ideally decreases by a factor of two by coupling two identical cores. In general, coupling N identical cores allows ideally reducing the phase noise by a factor N. Equivalently, it can be said that coupling N cores lowers the phase noise of $10{\cdot}\log(N)$ dBc/Hz. Unfortunately, at the same time, the multi-core technique causes an increment by factor N of both power consumption and silicon area occupation, compared to a single oscillator core, if special arrangements are not adopted. Moreover, circuit and layout complexities increase with the number of coupled cores, which can degrade the phase noise benefit due to parasitics. Therefore, a good tradeoff between complexity and phase noise reduction is represented by the quad-core solutions.

3. Interstacked Transformers

At mm-wave frequencies, integrated transformers suffer from very poor magnetic coupling factor, k, due to the low coil inductance values typically adopted (e.g., 50–150 pH). Moreover, interconnection parasitics further limit the magnetic coupling since their weight can represent up to 30% of the overall inductance. Integrated transformers are usually implemented by adopting conventional configurations, such as interleaved or stacked spirals as shown in Figures 1a and 1b,c, respectively, according to specific performance requirements and operating frequencies [23–27]. Traditional configurations present pros and cons. Interleaved transformers take advantage of multilayer symmetric spirals to maximize both primary and secondary Q-factors, but do not achieve high k. On the other hand, stacked transformers are area efficient and have better k, while losing the electrical symmetry between coils that is often mandatory in some circuits. Unfortunately, stacked transformers suffer greatly from magnetic coupling degradation at mm-wave frequencies (i.e., when sub-nH inductance values are used).

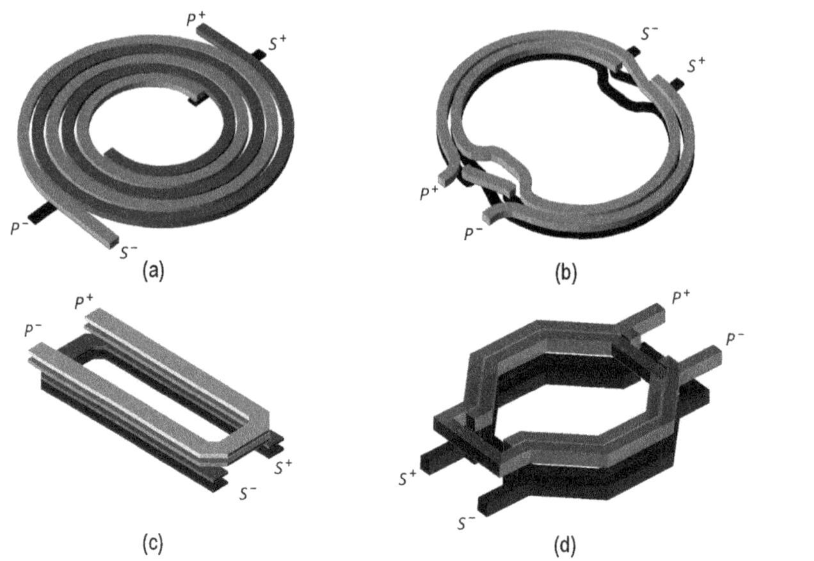

Figure 1. Transformer configurations: (**a**) Interleaved, (**b**) Stacked, (**c**) Folded stacked, (**d**) Interstacked.

The implementation of an area-efficient quad-core oscillator requires the availability of a high-k symmetric transformer configuration, as with the interstacked one shown in Figure 1d [28]. The interstacked transformer takes advantage of mixed interleaved/stacked windings, which improves the magnetic coupling between primary and secondary coils, still preserving geometrical/electrical symmetry and easiness of input/output connections. Specifically, it consists of two spirals of different metal layers, using complementary structures for primary and secondary windings. Indeed, the outer (inner) spiral of the primary winding is stacked to the outer (inner) spiral of the secondary winding and interleaved with the inner (outer) spiral of the secondary winding at the same time, thus exploiting both interleaved and stacked magnetic couplings. The interstacked transformer has several benefits at mm-wave frequencies. The improvement of magnetic coupling becomes significant in mm-wave optimized technologies that use thicker oxides between upper metal layers. In this case, the advantage of stacked coils is highly reduced and the interstacked structure can improve the overall magnetic coupling by more than 10%, especially when close inter-metal spacing is used [28]. Another important benefit of the interstacked structure is the electrical/geometrical symmetry, which provides similar performance for primary and secondary windings in terms of inductance and Q-factor.

Despite the advantages, the interstacked configuration has been rarely used due to higher design complexity compared to standard transformers. The interstacked transformer has been mainly exploited in improving the efficiency of the output matching network of a 77 GHz power amplifier [29].

As a first but effective design of an interstacked transformer, a simple lumped model has been proposed in [30]. For the sake of completeness, the model schematic is reported in Figure 2, along with main design equations and geometrical parameters. It is made up of two π-like networks, each for the primary and the secondary windings. The primary (secondary) winding is composed of a top MT (bottom MB) outer spiral shunted to a bottom

MB (top MT) inner spiral. As a consequence, the total inductance of the primary (secondary) transformer winding can be modeled by the shunt of two inductances, L_{MTP} and L_{MBP} (L_{MTS} and L_{MBS}), corresponding to the outer (inner) and inner (outer) spirals. To consider the positive mutual inductance between the spirals of the same winding, additional inductances, L_m, are also added in series. The low-frequency inductance values of outer (L_{OUT}) and inner (L_{IN}) spirals can be calculated by using the current sheet expression for octagonal coils (a) [31], where d_{avg} is the average diameter and ρ is the fill factor calculated according to equations (d) and (e). The series contribution, L_m, is drawn from the mutual coupling between spirals (by means of the magnetic coupling factor k_m). The model uses two series resistances in both primary and secondary windings corresponding to the outer and inner metal coils. Their frequency-dependent values are calculated using expression (b), where $R_{OUT,IN}$ and $R_{DCOUT,IN}$ are the overall and dc series resistances of both outer and inner spirals, respectively, and f is the frequency (expressed in GHz). $R_{DC\,OUT,IN}$ can be calculated using the geometrical parameters of the spiral and its metal sheet resistance.

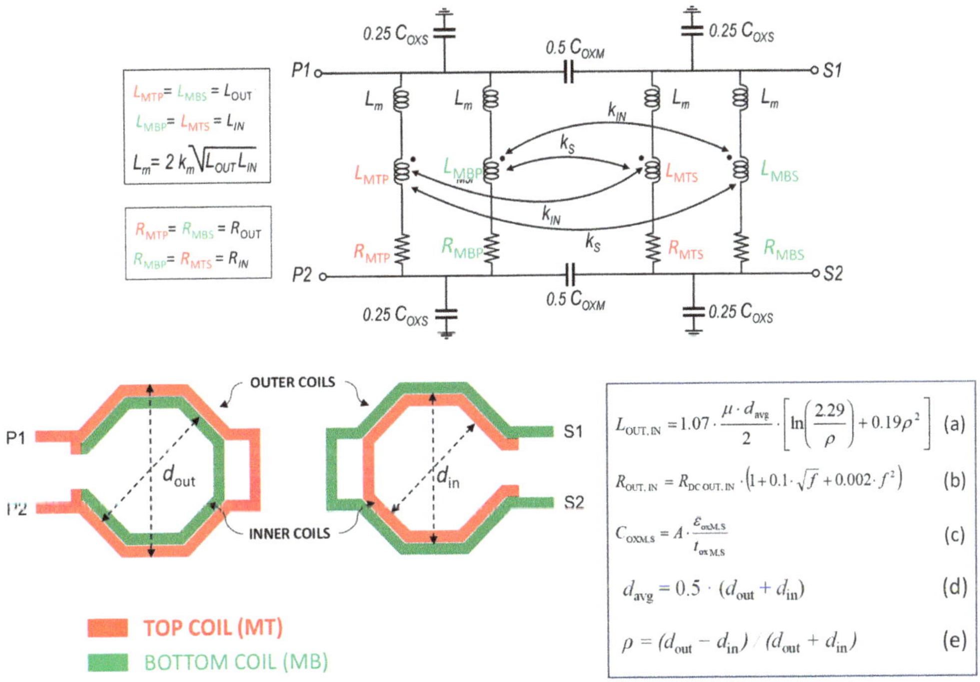

$$L_{OUT,IN} = 1.07 \cdot \frac{\mu \cdot d_{avg}}{2} \cdot \left[\ln\left(\frac{2.29}{\rho}\right) + 0.19\rho^2 \right] \quad \text{(a)}$$

$$R_{OUT,IN} = R_{DC\,OUT,IN} \cdot \left(1 + 0.1 \cdot \sqrt{f} + 0.002 \cdot f^2\right) \quad \text{(b)}$$

$$C_{OXM,S} = A \cdot \frac{\varepsilon_{oxM,S}}{t_{oxM,S}} \quad \text{(c)}$$

$$d_{avg} = 0.5 \cdot (d_{out} + d_{in}) \quad \text{(d)}$$

$$\rho = (d_{out} - d_{in}) / (d_{out} + d_{in}) \quad \text{(e)}$$

Figure 2. Physics-based scalable lumped model for interstacked transformers with related equations.

Magnetic coupling effects consist of two phenomena:

- The stacked coupling between outer (inner) spirals built in the top metal in primary and secondary windings, respectively, modeled by including a magnetic coupling factor, k_S, between L_{MTP} and L_{MBS} (and between L_{MBP} and L_{MTS}).
- The interleaved magnetic coupling between outer and inner spirals built in the same metal layer and belonging to different windings, modeled by adding a magnetic coupling factor, k_{IN}, between L_{MTP} and L_{MTS} (and between L_{MBP} and L_{MBS}).

Coupling factors k_m, $k_{S,}$, and k_{IN} could be drawn from EM simulations of open-air coupled spirals (i.e., without the silicon substrate beneath) separated by silicon dioxide of proper thickness, as proposed in [30]. Finally, capacitive effects are taken into account by means of the port-to-port inter-metal (C_{OXM}) and port-to-substrate (C_{OXS}) capacitances that mainly arise from the area contributions. They are calculated by using the expression (c), where $\varepsilon_{oxM,S}$ and $t_{oxM,S}$ are the oxide dielectric constant and thickness, respectively, and A is the area of the spirals. The model geometrical scalability allows using it as a simple but effective tool for the starting design of interstacked coils, while exploiting EM simulations for final accurate modeling.

4. Design of an Interstacked Four-Port Transformer

A reduction in the power consumption of VCOs can mainly be achieved by increasing the tank inductance value. This generally leads to designing larger and larger inductors, consequently increasing area occupation, and hindering the adoption of such solutions in commercial products. By adopting a transformer-based solution, it is possible to exploit the magnetic coupling between the coils to obtain large inductance values without impacting the area. Ideally, the greater the coupling factor, k, between the coils, the greater the area reduction and the better the coupling between the cores. Interstacked transformers—which exploit both the interleaved and stacked magnetic coupling between the coils—have proven to be a good solution for achieving high k values [28–30]. Compared to commonly used stacked transformers, they provide comparable or even higher *k-factors* and are intrinsically symmetrical. It should be noted that it is hard to find already-available solutions for interstacked four-port transformers. The main challenge of such a structure is to design a layout such that four coils—one for each core—can be interstacked by using only two metal layers, which is something that cannot be achieved by a simple stacked-transformer solution.

A 28 nm fully depleted silicon-on-insulator (FDSOI) CMOS technology is used in this work. It provides a BEOL having a thicker aluminum upper layer (ALUCAP) and two thinner copper lower layers, used in shunt—for the proposed transformer design—to minimize the difference in thickness between lower and upper metal layers. A simplified BEOL of the adopted CMOS technology is shown in Figure 3.

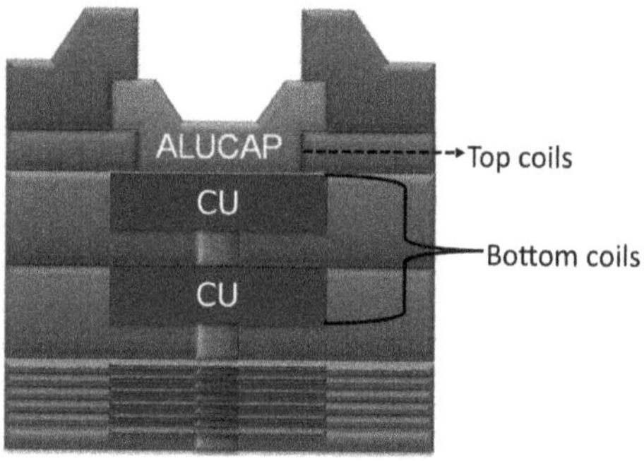

Figure 3. BEOL of the adopted 28 nm CMOS technology with eight metal layers [15,32].

The novel interstacked four-port transformer implementation is shown in Figure 4a. No patterned ground shield (PGS) was implemented to improve the self-resonance frequency (SRF) [33,34]. To better understand the geometry of the transformer, Figure 4b depicts the single coil which the whole structure is composed of. This "*squashed*" coil is obtained by the composition of two octagonal coils having different diameters, as shown in Figure 4c. Four of these coils are overlapped—each rotated by 90 degrees with respect to each other—to build the four-port structure of Figure 4a. An important aspect to mention is that there is a geometrical constraint that correlates the width of the coils and the inner diameter of the structure. For a given inner diameter, d_{in}, and for this given technology, the maximum coil width, w_{max}, that cannot be exceeded is given by the following expression.

$$w_{max} \approx 0.41 \cdot d_{inner} - 15.6\ \mu m \tag{1}$$

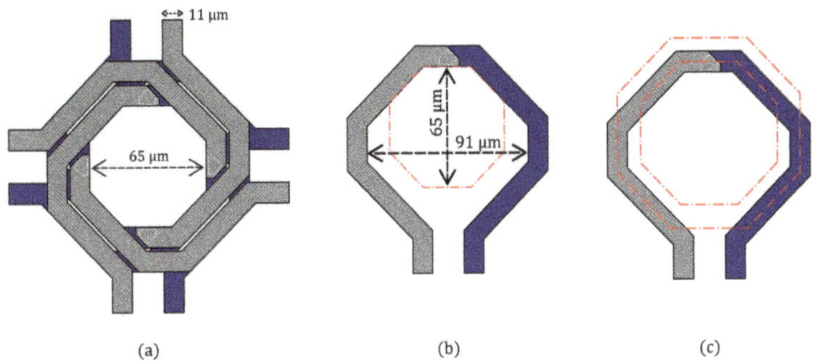

Figure 4. (**a**) Interstacked four-port transformer, (**b**) one of the four *"squashed"* coils used to build the interstacked transformer, (**c**) two octagons—the red dashed–dotted line—with different diameters are overlapped to the coil to highlight its geometry.

If a greater coil width was chosen, it would be impossible to guarantee the minimum spacing between the metal traces belonging to the same layer. Equation (1) suggests that there is a tradeoff between the coil width—which is proportional to the *Q*-factor—and how low the inductance can be. A coil width, *w*, of 11 μm was chosen for the proposed design, leading to an inner diameter of 65 μm. Table 1 reports the geometric data of the proposed transformer. A single coil, such as the one depicted in Figure 4b and having the size shown in Table 1, has an inductance of 146 pH and a *Q*-factor of 22 at 19.125 GHz. However, thanks to the magnetic coupling between the coils of the interstacked transformer, the equivalent inductance seen by each core is much higher than the one of a single coil, while keeping the same area occupation. Figure 5 shows the magnetic coupling factors between the four coils of the interstacked transformer. It is worth noting that slightly higher values are found for the magnetic coupling factors between orthogonal coils, given a slightly greater overlapping between them, as shown in Figure 6a,b. However, at the operating frequency of 19.125 GHz, *k* is always higher than 0.5; that is a very good result. Indeed, very small (symmetric) coils are used, which further confirms the advantages of the adopted interstacked structure.

Table 1. Geometric parameters of the proposed interstacked transformer.

Parameter	Value	Unit
Coil width	11	[μm]
Inner diameter [1], d_{IN}	65	[μm]
Outer diameter [1], d_{OUT}	91	[μm]
X-size	157	[μm]
Y-size	157	[μm]
Area	0.025	[mm^2]

[1] Inner and outer diameters are referred to Figure 4b.

Figure 7 reports the equivalent inductance seen by each port and its corresponding *Q*-factor. A value of 363 pH at 19.125 GHz is achieved. Its corresponding *Q*-factor at 19.125 GHz is as high as 21. The SRF is 69 GHz. It is worth mentioning that by coupling the four cores, the tank impedance of each one of them is in shunt with the others. The total effective tank impedance is then a quarter of the one seen by each port. That means that the total effective inductance at 19.125 GHz is as low as 90.8 pH.

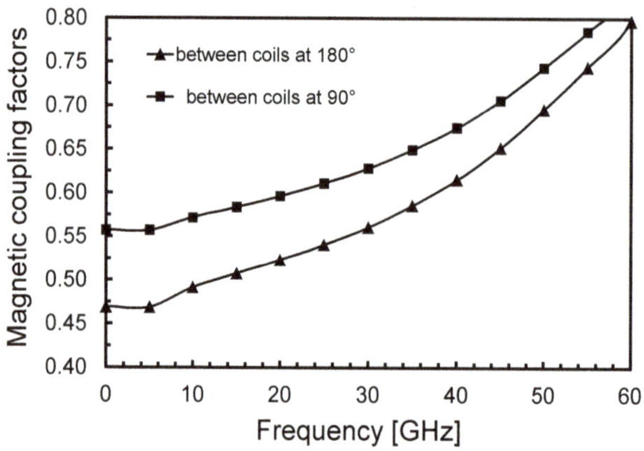

Figure 5. Magnetic coupling factors between the coils of the interstacked transformers.

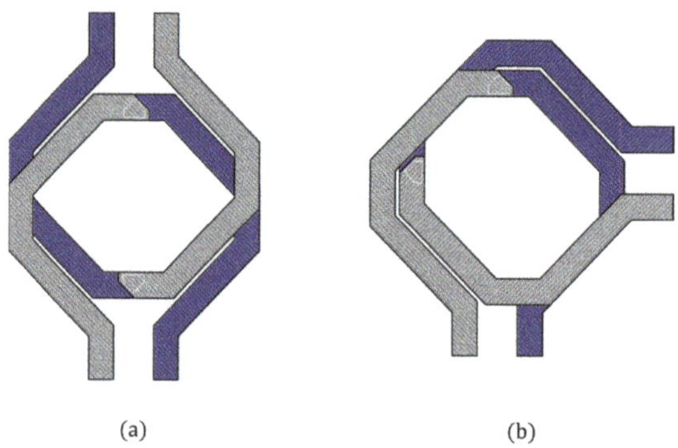

Figure 6. View of the overlapping between (**a**) opposite coils (180°) and (**b**) orthogonal coils (90°).

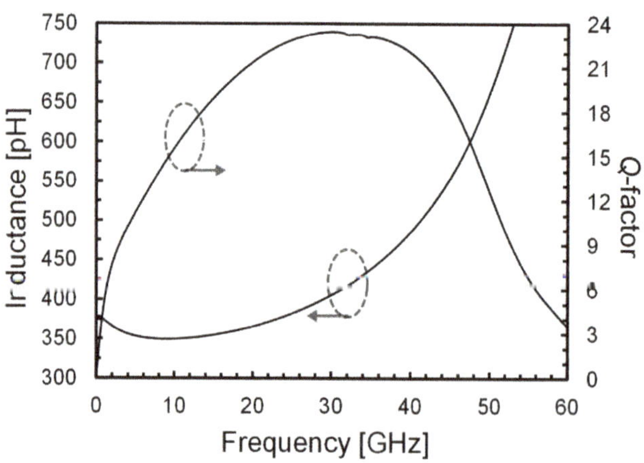

Figure 7. Equivalent inductance and Q-factor seen by each port of the interstacked transformer.

Since the interstacked transformer has four ports, four different oscillation modes could be triggered at startup. A good transformer design allows selecting only the desired mode and suppressing the others. This is achieved if the inductance has a high Q-factor at the desired oscillation mode, while being low at the undesired ones. A further simulation was then carried out to find all four possible oscillation modes with relative Q-factor. The results are reported in Table 2. The proposed solution is quite robust in this regard. Only the desired mode (mode 1) has a high Q, because of the constructive magnetic coupling between the coils. All other modes—which are found at higher frequencies—have small Q-factors due to the destructive coupling. Oscillation occurs at the desired mode at the startup. This analysis suggests that a strong magnetic coupling between the coils—such as the one provided by an interstacked transformer—carries the advantage of suppressing all the undesired modes, leading to no mode ambiguity, which is one of the most critical issues of multi-core oscillators.

Table 2. Oscillation modes for the interstacked transformer and corresponding inductance and Q-factor values.

Mode	Inductance (pH)	Q-Factor
1 (desired)	363	21.0
2 (undesired)	71.2	6.15
3 (undesired)	44.6	6.06
4 (undesired)	64.5	5.84

5. Quad-Core VCO Design Based on an Interstacked Transformer

A quad-core VCO to be used in a narrowband 77 GHz CMOS radar system for automotive applications [15] was designed in 28 nm fully depleted silicon on insulator (FD-SOI) CMOS technology by STMicroelectronics, exploiting the four-port interstacked transformer shown in Section 4. The technology provides low-V_T transistors, which exhibit a transition frequency, f_T, up to 270 GHz [32]. The process features a general purpose BEOL with eight copper layers and a top aluminum one [15], as shown in Figure 3.

Figure 8 reports a complete schematic of the quad-core VCO. The operating frequency of the VCO was set at 19.125 GHz to be later multiplied by four along the radar chain. A complementary cross-coupled oscillator was used since the supply cannot be fed through a center tap in the transformer. Moreover, no tail current was used, to maximize the output voltage swing. A supply voltage, V_{DD}, of 1.2 V was chosen. The transconductance, g_m, of the cross-coupled transistors was set at three times the minimum required by the Barkhausen criterion to obtain a robust startup and to achieve an oscillation amplitude of about 1 V. For the PMOS and the NMOS transistors, 63 nm and 90 nm transistors' channel lengths, respectively, were used to minimize the transistors' flicker and thermal noise, without excessively increasing their parasitic capacitances. The ratio $(W/L)_P/(W/L)_N$ was chosen to set the dc output voltage at $V_{DD}/2$. Finally, a differential varactor provides a 5.5% tuning range (around 1 GHz), covering the operating frequency range, the temperature variations, and part of the process variations. To be able to fully cover process variations, a 3-bit switched capacitor bank is exploited, achieving a total of 20% tuning range. A minimum overlap of 460 MHz between the bands is guaranteed.

The achieved performance, along with a comparison with the state-of-the-art quad-core solutions, is reported in Table 3. The expressions of the figures of merit (FoMs) used for the comparison are the following [20–22,35]:

$$\mathrm{FoM} = \mathrm{PN} - 20\log_{10}\left(\frac{f_0}{\Delta f}\right) + 10\log_{10}(P_{\mathrm{DC,mW}}), \quad (2)$$

$$\mathrm{FoM_T} = \mathrm{FoM} - 20\log_{10}\left(\frac{\mathrm{TR}}{10}\right), \quad (3)$$

$$\mathrm{FoM_A} = \mathrm{FoM} + 10\log_{10}(\mathrm{Area_{mm^2}}), \quad (4)$$

where PN is the phase noise, f_0 is the oscillation frequency, Δf is the offset frequency from the carrier at which phase noise is measured, $P_{DC,mW}$ is the dc power consumption expressed in mW, TR is the frequency tuning range, and $Area_{mm}{}^2$ is the area consumption expressed in mm^2. The oscillator phase noise is shown in Figure 9.

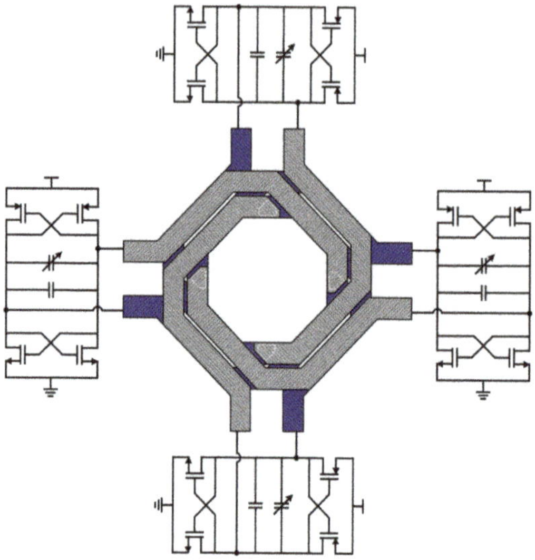

Figure 8. Schematic of the proposed quad-core VCO with interstacked four-port transformer.

Table 3. Performance comparison with the state-of-the-art quad-core VCOs.

	[20]	[21]	[22]	This Work
Inductor topology	Single-turn	Four-port	Stacked	Interstacked
VCO topology	Class-B, NMOS-only, tail filtering	Complementary cross-coupled, $4f_0$tail filtering	NMOS-only, tail filtering, push-push	Complementary cross-coupled
Technology	55 nm BiCMOS	40 nm CMOS	45 nm PDSOI CMOS	28 nm FDSOI CMOS
Frequency (GHz)	20	26.45	60.5	19.125
Tuning range (%)	15	26	19	20
Tuning range type	analog and discrete	analog and discrete	analog-only	analog and discrete
PN @ 1 MHz (dBc/Hz)	−118.5	−109.5	−101.7	−112.6
PN @ 1 MHz from 19.125 GHz (dBc/Hz)	−118.9	−112.3	−111.7	−112.6
V_{DD} (V)	1.2	0.95	1	1.2
Current (mA)	36	17	40	6
Power (mW)	43	16	40	7.2
Area (mm^2)	0.6	0.1	0.044	0.025
FoM (dBc/Hz)	−188.2	−186.8	−181.3	−189.6
FoM_T (dBc/Hz)	−191.7	−195.0	−186.9	−195.6
FoM_A (dBc/Hz)	−190.4	−196.8	−194.9	−205.6

The proposed interstacked transformer quad-core VCO design exhibits a low phase noise while having minimum power consumption and area occupation. It achieves a 24× area and 6× power consumption reduction with respect to [20]. It has 1 dB lower phase noise than the stacked-transformer-based solution [22], along with a 7× reduction in power consumption and a 43% reduction in area occupation.

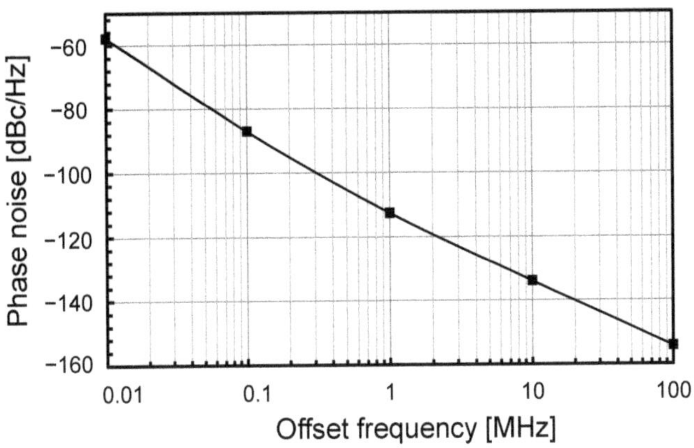

Figure 9. Interstacked VCO phase noise at 1 MHz offset from the carrier.

6. Conclusions

A quad-core VCO, exploiting an interstacked four-port transformer to magnetically couple the four cores, has been proposed. As expected, area occupation and power consumption are extremely low, while keeping the phase noise at a low level. This allows its use in commercial products. The comparison versus state-of-the-art quad-core VCOs further confirms the strength of the proposed solution. Although the focus of the present work is 77 GHz radar, the field of application is wide and includes all modern wireless communication systems.

Author Contributions: Conceptualization, D.T., G.M. and E.R.; validation, D.T. and G.M.; formal analysis, D.T.; methodology, D.T.; project administration, E.R.; supervision, S.C.P. and E.R.; writing—original draft, D.T. and E.R.; writing—review and editing, D.T., E.R. and S.C.P. All authors have read and agreed to the published version of the manuscript.

Funding: This work was supported by the European Union under the Italian National Recovery and Resilience Plan (PNRR) of Next Generation EU, partnership on "Telecommunications of the Future" (PE0000001—program "RESTART").

Data Availability Statement: The data presented in this study are available on request from the corresponding author.

Conflicts of Interest: Authors Daniele Tripoli and Giorgio Maiellaro were employed by the company STMicroelectronics. The remaining authors declare that the research was conducted in the absence of any commercial or financial relationships that could be construed as a potential conflict of interest.

References

1. Siddiq, K.; Watson, R.J.; Pennock, S.R.; Avery, P.; Poulton, R.; Dakin-Norris, B. Phase noise analysis in FMCW radar systems. In Proceedings of the 2015 European Microwave Conference (EuMC), Paris, France, 7–10 September 2015; pp. 501–504.
2. Thurn, K.; Ebelt, R.; Vossiek, M. Noise in homodyne FMCW radar systems and its effects on ranging precision. In Proceedings of the 2013 IEEE MTT-S International Microwave Symposium Digest, Seattle, WA, USA, 2–7 June 2013; pp. 1–3.
3. Hajimiri, A.; Lee, T.H. Design issues in CMOS differential LC oscillators. *IEEE J. Solid-State Circuits* **1999**, *34*, 717–724. [CrossRef]
4. Krishnaswamy, H.; Hashemi, H. Inductor- and transformer-based integrated RF oscillators: A comparative study. In Proceedings of the IEEE Custom Integrated Circuits Conference, San Jose, CA, USA, 10–13 September 2006; pp. 381–384.
5. Mazzanti, A.; Bevilacqua, A. On the phase noise performance of transformer-based CMOS differential-pair harmonic oscillators. *IEEE Trans. Circuits Syst. I Regular Papers* **2015**, *62*, 2334–2341. [CrossRef]
6. Issakov, V.; Padovan, F. A dual-core 60 GHz push-push VCO with second harmonic extraction by mode separation. In Proceedings of the IEEE Radio Frequency Integrated Circuits Symposium (RFIC), Philadelphia, PA, USA, 10–12 June 2018; pp. 208–211.
7. Issakov, V.; Padovan, F.; Rimmelspacher, J.; Weige, R.; Geiselbrechtinger, A. A 52-to-61 GHz push-push VCO in 28 nm CMOS. In Proceedings of the European Microwave Conference (EuMC), Madrid, Spain, 23–27 September 2018; pp. 1009–1012.

8. Cavarra, A.; Papotto, G.; Parisi, A.; Finocchiaro, A.; Nocera, C.; Palmisano, G. Transformer-based VCO for W-band automotive radar applications. *Electronics* **2021**, *10*, 531. [CrossRef]
9. Trotta, S.; Wintermantel, M.; Dixon, J.; Moeller, U.; Jammers, R.; Hauck, T.; Samulak, A.; Dehlink, B.; Shun-Meen, K.; Li, H.; et al. An RCP packaged transceiver chipset for automotive LRR and SRR systems in SiGe BiCMOS technology. *IEEE Trans. Microw. Theory Tech.* **2012**, *60*, 778–794. [CrossRef]
10. Belfiore, F.; Calcagno, A.; Borgonovo, G.; Castro, M.G.; Pisasale, A.; Platania, M.; Vinciguerra, M.; Schiro, C.; Alessi, G.; Burgio, C.; et al. A 76 to 81 GHz packaged transceiver for automotive radar with FMCW modulator and ADC. In Proceedings of the 2017 European Radar Conference (EURAD), Nuremberg, Germany, 11–13 October 2017.
11. Lou, L.; Tang, K.; Chen, B.; Guo, T.; Wang, Y.; Wang, W.; Fang, Z.; Liu, Z.; Zheng, Y. A 253mW/channel 4TX/4RX pulsed chirping phased-array radar TRX in 65nm CMOS for X-band synthetic-aperture radar imaging. In Proceedings of the 2018 IEEE International Solid-State Circuits Conference-(ISSCC), San Francisco, CA, USA, 11–15 February 2018; pp. 160–162.
12. Ma, T.; Deng, W.; Chen, Z.; Wu, J.; Zheng, W.; Wang, S.; Qi, N.; Liu, Y.; Chi, B. A CMOS 76–81-GHz 2-TX 3-RX FMCW radar transceiver based on mixed-mode PLL chirp generator. *IEEE J. Solid-State Circuits* **2020**, *55*, 233–248. [CrossRef]
13. Arai, T.; Usugi, T.; Murakami, T.; Kishimoto, S.; Utagawa, Y.; Kohtani, M.; Ando, I.; Matsunaga, K.; Arai, C.; Yamaura, S.; et al. A 77-GHz 8RX3TX transceiver for 250-m long-range automotive radar in 40-nm CMOS technology. *IEEE J. Solid-State Circuits* **2021**, *56*, 1332–1344. [CrossRef]
14. Papotto, G.; Parisi, A.; Finocchiaro, A.; Nocera, C.; Cavarra, A.; Castorina, A.; Palmisano, G. A W-Band transmitter for automotive radar sensors in 28-nm FD-SOI CMOS. *IEEE Trans. Microw. Theory Tech.* **2023**, *71*, 4577–4587. [CrossRef]
15. Ragonese, E.; Nocera, C.; Cavarra, A.; Papotto, G.; Spataro, S.; Palmisano, G. A comparative analysis between standard and mm-wave optimized BEOL in a nanoscale CMOS technology. *Electronics* **2020**, *9*, 2124. [CrossRef]
16. Joseph, A.; Jain, V.; Ong, S.N.; Wolf, R.; Lim, S.F.; Singh, J. Technology positioning for mm wave applications: 130/90 nm SiGe BiCMOS vs. 28 nm RFCMOS. In Proceedings of the IEEE BiCMOS and Compound Semiconductor Integrated Circuits and Technology Symposium (BCICTS), San Diego, CA, USA, 15–17 October 2018; pp. 18–21.
17. Ragonese, E.; Papotto, G.; Nocera, C.; Cavarra, A.; Palmisano, G. CMOS automotive radar sensors: Mm Wave circuit design challenges. *IEEE Trans. Circuits Syst. II Express Briefs* **2022**, *69*, 2610–2616. [CrossRef]
18. Ahmadi-Mehr, S.A.-R.; Tohidian, M.; Staszewski, R.B. Analysis and design of a multi-core oscillator for ultra-low phase noise. *IEEE Trans. Circuits Syst. I Regul. Pap.* **2016**, *63*, 529–539. [CrossRef]
19. Deng, Z.; Niknejad, A.M. A 4-port-inductor-based VCO coupling method for phase noise reduction. *IEEE J. Solid-State Circuits* **2011**, *46*, 1772–1781. [CrossRef]
20. Iotti, I.; Mazzanti, A.; Svelto, F. Insights into phase-noise scaling in switch-coupled multi-core LC VCOs for E-Band adaptive modulation links. *IEEE J. Solid-State Circuits* **2017**, *52*, 1703–1718. [CrossRef]
21. Murphy, D.; Darabi, H. A 27-GHz quad-core CMOS oscillator with no mode ambiguity. *IEEE J. Solid-State Circuits* **2018**, *53*, 3208–3216. [CrossRef]
22. Rimmelspacher, J.; Weigel, R.; Hagelauer, V.; Issakov, V. A Quad-core 60 GHz push-push 45 nm SOI CMOS VCO with −101.7 dBc/Hz phase noise at 1 MHz offset, 19% continuous FTR and −187 dBc/Hz FoMT. In Proceedings of the IEEE European Solid State Circuits Conference (ESSCIRC), Dresden, Germany, 3–6 September 2018; pp. 138–141.
23. Long, J.R. Monolithic transformers for silicon RF IC design. *IEEE J. Solid-State Circuits* **2000**, *35*, 1368–1382. [CrossRef]
24. Long, J.R. On-chip transformer design and application to RF and mm-wave front-ends. In Proceedings of the IEEE Custom Integrated Circuits Conference (CICC), Austin, TX, USA, 3 May 2017; pp. 1–43.
25. Rimmelspacher, J.; Breun, S.; Werthof, A.; Geiselbrechtinger, A.; Weigel, R.; Issakov, V. Experimental comparison of integrated transformers in a 28 nm bulk CMOS technology. In Proceedings of the European Microwave Conference (EuMC), Madrid, Spain, 23–27 September 2018; pp. 1097–1100.
26. Spataro, S.; Ragonese, E. Design and optimization of silicon-integrated inductive components for automotive radar applications in K- and W-bands. In Proceedings of the 2020 AEIT International Conference of Electrical and Electronic Technologies for Automotive (AEIT AUTOMOTIVE), Torino, Italy, 18–20 November 2020; pp. 1–6.
27. Bevilacqua, A. Fundamentals of integrated transformers: From principles to applications. *IEEE Solid-State Circuits Mag.* **2020**, *12*, 86–100. [CrossRef]
28. Ragonese, E.; Sapone, G.; Palmisano, G. High-performance interstacked transformers for mm-wave ICs. *Microw. Opt. Technol. Lett.* **2010**, *52*, 2160–2163. [CrossRef]
29. Giammello, V.; Ragonese, E.; Palmisano, G. A transformer-coupling current reuse SiGe HBT power amplifier for 77-GHz automotive radar. *IEEE Trans. Microw. Theory Tech.* **2012**, *60*, 1676–1683. [CrossRef]
30. Ragonese, E.; Sapone, G.; Giammello, V.; Palmisano, G. Analysis and modeling of interstacked transformers for mm-wave applications. *Analog. Integr. Circuits Signal Process.* **2012**, *72*, 121–128. [CrossRef]
31. Mohan, S.S.; del Mar Hershenson, M.; Boyd, S.P.; Lee, T.H. Simple accurate expressions for planar spiral inductances. *IEEE J. Solid-State Circuits* **1999**, *34*, 1419–1424. [CrossRef]
32. Cathelin, A. Fully depleted silicon on insulator devices CMOS: The 28-nm node is the perfect technology for analog, RF, mmW, and mixed-signal system-on-chip integration. *IEEE Solid-State Circuits Mag.* **2017**, *9*, 18–26. [CrossRef]

33. Liang, H.-B.; Lin, Y.-S.; Chen, C.-C.; Lee, J.-H. Optimization of PGS pattern of transformers/inductors in standard RF BiCMOS technology for RFIC applications. In Proceedings of the IEEE Radio Frequency Integrated Circuits (RFIC) Symposium, San Francisco, CA, USA, 10–13 June 2006; p. 4.
34. Spataro, S.; Salerno, N.; Papotto, G.; Ragonese, E. The effect of a metal PGS on the Q-factor of spiral inductors for RF and mm-wave applications in a 28-nm CMOS technology. *Int. J. RF Microw. Comput.-Aided Eng.* **2020**, *30*, e22368. [CrossRef]
35. Mansour, I.; Aboualalaa, M.; Barakat, A.; Allam, A.; Abdel-Rahman, A.B.; Abo-Zahhad, M.; Pokharel, R.K. Analysis and implementation of high-Q CT inductor for compact and wide-tuning range Ku-band VCO. *IEEE Microw. Wirel. Compon. Lett.* **2020**, *30*, 802–805. [CrossRef]

electronics

MDPI

Article

A Miniaturized Bandpass Filter with Wideband and High Stopband Rejection Using LTCC Technology

Yue Ma [1,2,3,4], Qifei Du [1,2,3,4,*], Wei Zhang [5], Cheng Liu [1,3,4] and Hao Zhang [1,3,4]

1 National Space Science Center, Beijing 100190, China; mayue21@mails.ucas.ac.cn (Y.M.); lc@nssc.ac.cn (C.L.); zhanghao@nssc.ac.cn (H.Z.)
2 University of Chinese Academy of Sciences, Beijing 100190, China
3 Beijing Key Laboratory of Space Environment Exploration, Beijing 100190, China
4 Key Laboratory of Science and Technology on Environmental Space Situation Awareness, Chinese Academy of Sciences, Beijing 100190, China
5 Shanghai Institute of Satellite Engineering, Shanghai 200240, China; wzhang509@126.com
* Correspondence: dqf@nssc.ac.cn

Abstract: This paper designs an L-band wide stopband bandpass filter by applying low-temperature cofired ceramic (LTCC) technology to the global positioning system (GPS) frequency band. Taking the Chebyshev filter as a prototype, an equivalent collector element (capacitive and inductor) structure is adopted to fully use the three-dimensional package structure of LTCC to reduce the filter size. The filter is integrated into an eight-layer LTCC dielectric, and the series–parallel connection of the collector elements in the resonance unit is utilized to produce out-of-band transmission zeros, while the input and output ports' capacitance is adjusted to control the bandwidth. Harmonic suppression is achieved by cascading two new compact stopband filters, while the size increase is insignificant due to LTCC technology. The simulation results are as follows: the center frequency is 1.575 GHz, 1 dB relative bandwidth is 6.3%, insertion loss in the passband is as slight as 1.6 dB, return loss is better than 30 dB, rejection bandwidth up to 16 GHz is more than 44 dB, and the volume of the whole filter is $6.2 \times 3.7 \times 0.78$ mm^3.

Keywords: L-band; LTCC technology; wide stopband; narrow passband; lumped elements

1. Introduction

Microwave filter is a crucial part of advanced wireless communication systems. Many microwave filters can be utilized to filter out the clutter harmonics. In recent years, given the system miniaturization requirements, especially passive component miniaturization requirements, low-temperature cofired ceramic (LTCC) technology as a new form of circuit implementation has been developed very quickly. Due to the high integration and high performance of LTCC technology, its three-dimensional encapsulation structure can be buried in the multilayer ceramic substrate passive components, thus reducing the size of the passive components to a large extent in the design of the system miniaturization [1]. Nowadays, various studies have been performed on microwave filter miniaturization, like the stepped-impedance resonator unit [2], the double-layer suspended stripline resonator unit [3], and the spiral resonator unit [4], which can reduce the resonator unit size to 1/4 wavelength. However, these resonant cells have a significant L-band size [5]. The LTCC three-dimensional structure has been employed to implement the lumped capacitive–inductive [6] as a stack in the Z-direction. The capacitive–inductive coupling is implemented between different layers, which can decrease the size of the microwave filters and produce transmission zeros (TZs) and cascaded bandstop filters. Thus, a filter with better attenuation performance is achieved.

This paper designs an L-band wide stopband bandpass filter using LTCC technology for the GPS frequency band. The filter comprises a narrow passband filter and two bandstop filters. The narrow filter generates transmission zeros (TZs) by introducing inductors based on a filter prototype containing an inverting converter. Compared to the existing narrowband filter prototype [7], the new filter prototype has TZs on the right side of the center frequency. The center frequency is 1.575 GHz, and the 1 dB relative bandwidth is 6.3%. The narrowband filters are cascaded with two bandstop filters to design a new bandstop filter. The resultant filter has three zeros in the high rejection band. It provides excellent blocking characteristics, achieving a wide bandwidth of 15.6 GHz, with a blocking rejection exceeding 44 dB. Compared with the recently developed wide stopband filters [8–10], this filter provides better stopband characteristics and a narrower passband. In addition, the size of this miniaturized narrowband filter is only 6.2 mm × 3.7 mm × 0.78 mm.

The remainder of this paper is structured as follows. The prototype circuit of the narrowband filter is introduced in Section 2. Section 3 presents the design of the three-dimensional structure of the filter. A novel bandstop filter structure is presented in Section 4. Finally, the simulation results after combining the narrowband filter with the novel bandstop filter are presented in Section 5. The advantages of the proposed filter are also described compared with the existing filters.

2. Filter Structure Design

This paper employs the proposed filter to fabricate the LC collector element filter using the LTCC process by fabricating the capacitors and inductors separately and embedding them in LTCC dielectric to realize the circuit [11]. Each component is manufactured using FerroA6M material from Ferro with a 5.9 dielectric constant, a loss angle tangent of 0.002, and a dielectric substrate thickness of 97 µm per layer (after cofiring) [12]. The conductor metal is gold, and the thickness of the metal line is 8 µm. The line design specifications, such as minimum line width, minimum line spacing, and through-hole diameter, are strictly compatible with those of the LTCC. Table 1 presents the particular efficiency requirements of the constructed bandpass filter.

Table 1. The filter's estimated indicators.

Feature	Value
1 dB passband range	1.52–1.62 GHz
Insertion loss	≤1.65 dB
Out-of-band rejection	6.2f0 ≥ 44 dB
Return loss	>30 dB
Dimension	0.036λ × 0.022λ

2.1. Filter Principle

As presented in Table 1's metrics, since the bandwidth requirements are narrow, an inverting converter is chosen to realize the bandpass filter design [13]. The inverting converter can convert the prototype shown in Figure 1, including both inductors and capacitors, into an equivalent form with only inductors or capacitors. Figure 2 [14] describes the inverting converter.

As shown in Figure 2a, an ideal input impedance Z_a inversion converter acts like a quarter-wavelength transmission line with a characteristic impedance of K at each frequency. That is, assuming an impedance is connected at one terminal, the impedance seen at the other terminal is as follows:

$$Z_a = \frac{K_{k,k+1}^2}{Z_b} \quad (1)$$

In the same way, in Figure 2b, an ideal conductance inverting converter acts similar to a quarter-wavelength transmission line containing a characteristic conductance of J at each frequency, such that the conductance is connected to one terminal, and the conductance seen from the other terminal is as follows:

$$Y_a = \frac{J_{k,k+1}^2}{Y_b} \quad (2)$$

Also, as shown in Figure 2, the mirror phase shift of the inverting converter is an odd multiple of plus or minus 90 degrees.

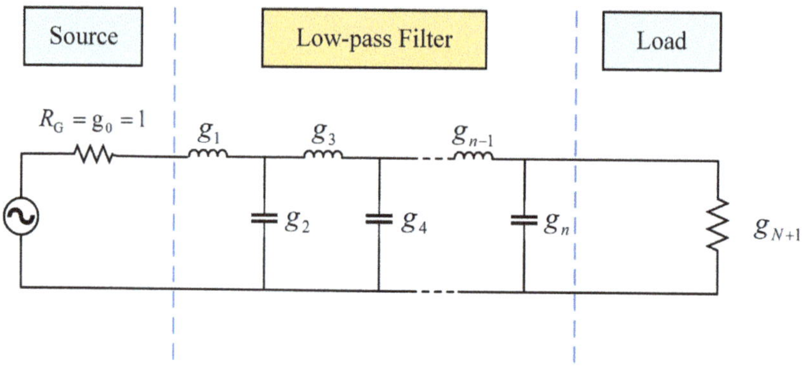

Figure 1. Low-pass filter prototype circuit.

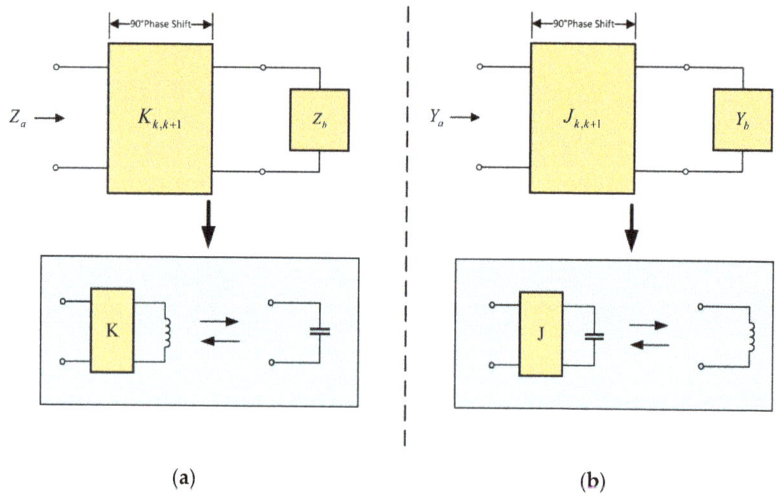

Figure 2. (**a**) Definition of an impedance inversion converter. (**b**) Definition of a conductance inversion converter.

The prototype low-pass filter circuit in Figure 1 can be transformed into two equivalent forms using the nature of the conductance inversion converter, as shown in Figure 3, including the same transmission characteristics as the low-pass prototype shown in Figure 1. Figure 3 shows a modified prototype circuit using an impedance inversion converter, where the values of G_A, G_B, and C_{rk} (C_{r1},...,C_{rn}) can be taken arbitrarily. When the parameter

$J_{k,k+1}$ of the inverting converter is obtained from Equations (3)–(5), the response will be the same as that of the filter prototype in Figure 1.

$$J_{01} = \sqrt{\frac{G_A C_{r1}}{g_0 g_1}} \quad (3)$$

$$J_{k,k+1|k=1\cdots n-1} = \sqrt{\frac{C_{rk} C_{r(k+1)}}{g_k g_{k+1}}} \quad (4)$$

$$J_{n,n+1} = \sqrt{\frac{C_{an} G_B}{g_n g_{n+1}}} \quad (5)$$

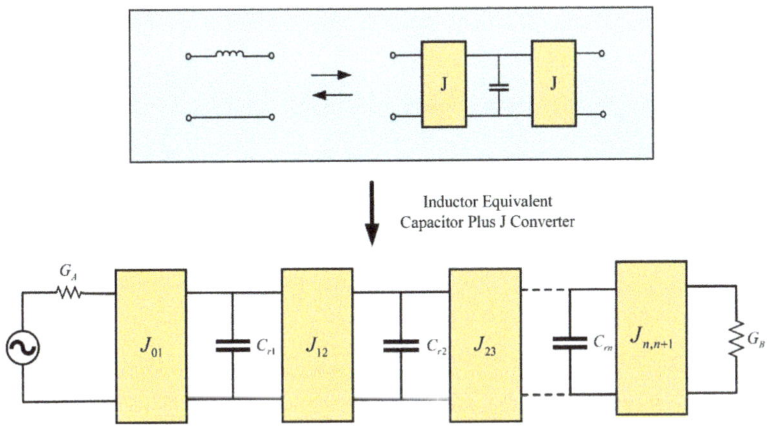

Figure 3. Modified low-pass prototype, including an impedance/conductance inverting converter.

It is essential to emphasize that g_k in the equation shown in Figure 3 (k = 0, 1, 2, ..., n + 1) is the component values of the low-pass prototype circuit described in Figure 1.

Exerting the low-pass filter to bandpass conversion on the circuit shown in Figure 3 gives a generalized bandpass filter with a conductive inverting converter and a shunt-type resonator, as shown in Figure 4.

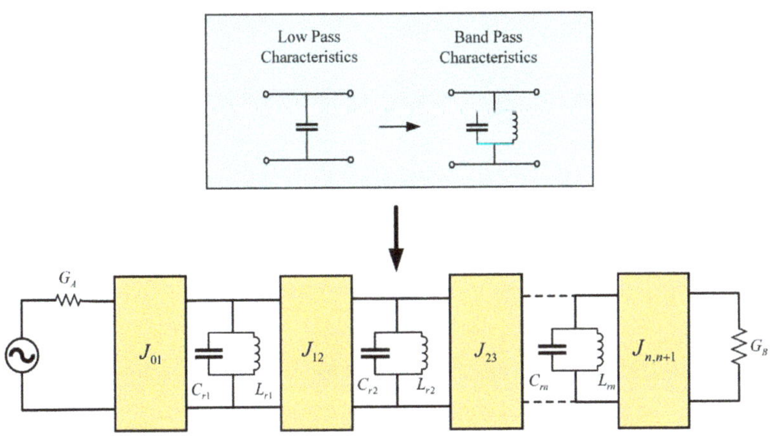

Figure 4. Coupled resonator bandpass filter transformed from Figure 3.

2.2. Filter Circuit Prototype

An inverting converter with aggregate components can be realized with different forms, such as inductive and capacitive forms, constructed with π-type and L-type structures. Among them, the capacitive form is chosen in this paper. Figure 5 shows the equivalent circuit of the above four inverting converters. It should be noted that a negative inductor (capacitor) in the circuit must be absorbed by a nearby positive inductor (capacitor), which turns out to be all positive components. Therefore, an L-structure capacitor converter is applied to both the first and last J-converters of the circuit and a π-structure capacitor converter is chosen in the middle to maintain a symmetrical structure. Taking the J inverting converter in Figure 4 as the form shown in Figure 5, a standard and specific implementation of Figure 4 can be attained using the generalized bandpass filter circuit for the conductive inverting converter mentioned in Section 2.1, as shown in Figure 6.

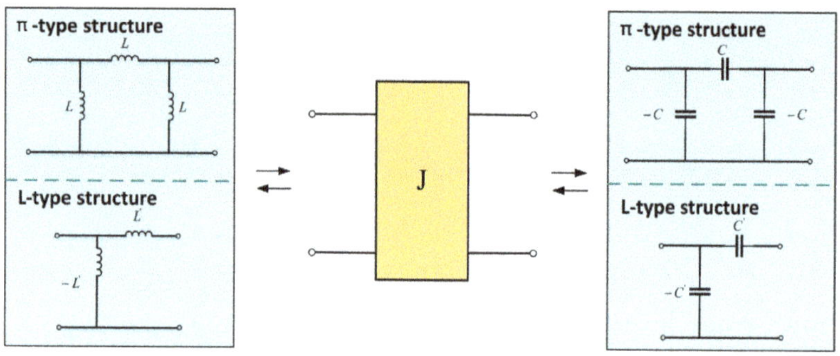

Figure 5. Specific circuit of the J inverting converter.

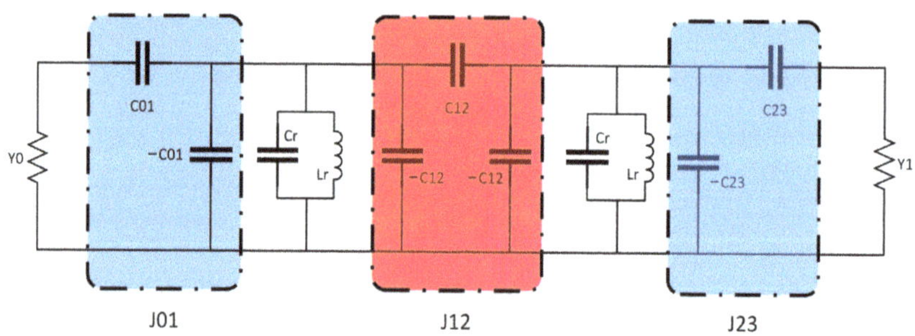

Figure 6. Transformed bandpass filter circuit.

The following formulas can determine the components of the specific circuit presented in Figure 6:

$$C_r = \frac{g}{2\pi \cdot BW \cdot Z_c} \tag{6}$$

$$L_r = \frac{2\pi \cdot BW \cdot Z_c}{g \cdot \omega_0^2} \tag{7}$$

$$b_i = \omega_0 \cdot C_{ri} \tag{8}$$

$$J_{01} = \sqrt{\frac{Y_0 \omega_0 b_1}{g_0 g_1}} \tag{9}$$

$$J_{12} = \omega_0 \sqrt{\frac{b_1 b_2}{g_1 g_2}} \tag{10}$$

$$J_{23} = \sqrt{\frac{Y_3 \omega_0 b_2}{g_2 g_3}} \tag{11}$$

$$C'_{01} = \frac{J_{0,1}}{\omega_0 \sqrt{1 - \left(\frac{J_{0,1}}{Y_0}\right)^2}} \tag{12}$$

$$C_{01} = \frac{C'_{01}}{1 + \left(\frac{\omega_0 \cdot C_{01}}{Y_0}\right)^2} \tag{13}$$

$$C'_{23} = \frac{J_{2,3}}{\omega_0 \sqrt{1 - \left(\frac{J_{2,3}}{Y_3}\right)^2}} \tag{14}$$

$$C_{23} = \frac{C'_{23}}{1 + \left(\frac{\omega_0 \cdot C_{23}}{Y_3}\right)^2} \tag{15}$$

where ω_0 is the passband center frequency, $Z_c = \frac{1}{Y_{i(i=0,3)}}$ is the characteristic impedance, g is the adjustment factor, which dictates the selection of a suitable capacitor and inductor, and BW denotes the 1 dB bandwidth. C'_{01} and C'_{23} are just intermediate values. $g_{i(i=0,1,2,3)}$ are Chebyshev values, chosen as 1, 0.843, 0.622, and 1.3554, respectively.

Based on a conductive inverted bandpass filter, an inductor L2 is added to introduce a deeper transmission zero in the filter. The Chebyshev type is modified to promote the sideband rejection. The filter's equivalent schematic is designed using the ADS software and is presented in Figure 7.

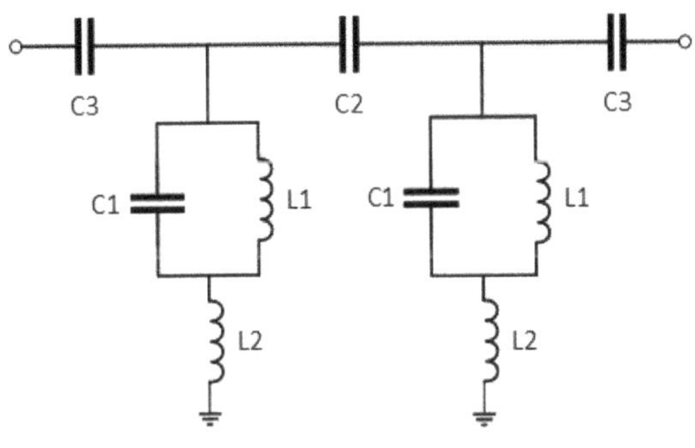

Figure 7. Narrowband filter schematic diagram.

The filter operating center frequency is 1.575 GHz, the 1 dB relative bandwidth is 100 MHz, and the initial values of the components in the equivalent circuit are obtained from Equations (6)–(12), as follows: C3 = 0.506 pF, C2 = 0.139 pF, C1 = 7.163 pF, and L1 = 1.283 nH. The value of L2 is derived directly from simulations, and the indices require wider stopband rejection. Therefore, adjusting the inductance L2 will attenuate the zero point near 6 GHz. The simulations indicate that the capacitances C2 and C3 adjust the bandwidth, C1 and L1 can adjust the passband center frequency, and L2 can adjust the TZ in Figures 8–10. As shown in Figure 8, when C2 and C3 increase, the filter bandwidth increases, and the insertion loss decreases. The filter bandwidth cannot be too narrow because it is limited by the insertion loss. Our target indicator's bandwidth and insertion loss are obtained by optimally selecting appropriate values of C2 and C3. Figure 9 shows that the center frequency of the filter passband can be adjusted very simply by adjusting the value of the inductor L1, and the same function can be achieved by adjusting the value of the capacitor C1. This paper aims to select the L1 frequency point of the GPS, thus achieving the L2 and L5 frequency points of the GPS. Figure 10 indicates that by adjusting the value of inductor L2, the high-frequency transmission zero point can be adjusted without changing the performance of the filter passband, thus attaining a good suppression effect on the high-frequency signal. This model achieves better suppression of high-frequency signals compared with the existing models, in which both of their transmission zeros are to the right of the center frequency [7]. When the two L2 inductances are equal, only one transmission zero can be generated, while the depression becomes deeper, otherwise, they will be splatted into two transmission zeros with a shallower depression depth. The former is chosen for higher suppression requirements.

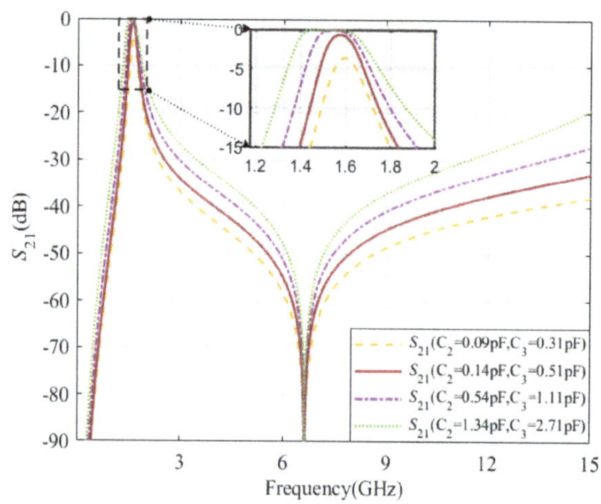

Figure 8. Effect of C2 and C3 on the filter's passband bandwidth and insertion loss.

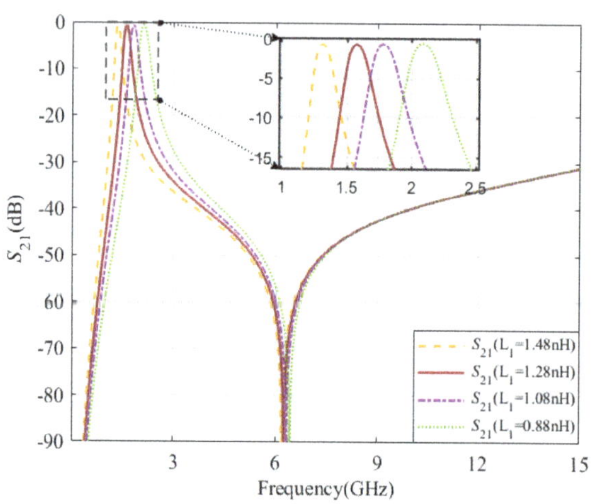

Figure 9. Effect of L1 on the filter's passband center frequency.

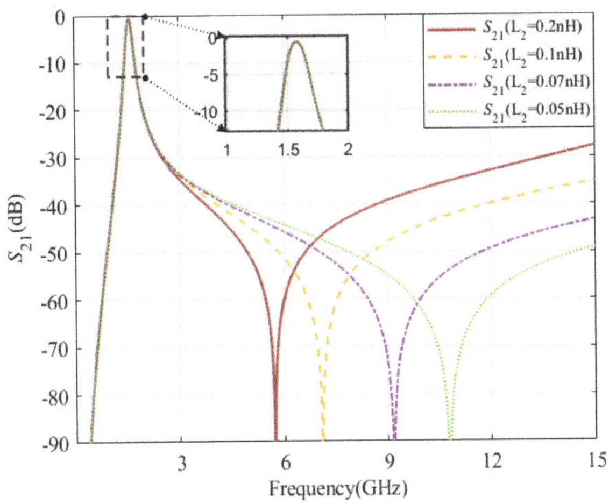

Figure 10. Effect of L2 on the filter's high-frequency transmission zeros.

3. 3D Structural Design

As the needed filter is a narrow band one, every slight adjustment to the capacitance and inductance values may influence the filter's bandwidth. Mutual coupling is necessary for the final realization of the filter using the LTCC technique. Therefore, 3D EM simulation with HFSS software is crucial in the current study [15]. The frequency model is solved using the adaptive meshing and broadband to obtain more accurate results in broadband simulations.

3.1. Capacitors

There are two main LTCC capacitors: mental insulator mental (MIM) capacitor and vertically interdigitated capacitor (VIC). VICs are the first choice for realizing capacitors with large capacitance, which can make full use of the LTCC technology's multilayer feature to increase the capacitance by increasing the number of layers. Table 2 compares the main performance and parameters of the two structures. As presented in Table 2, under the same effective capacitance value, the VIC structure has a higher self-resonant frequency (SRF), higher quality factor (Q), and a small area required for each layer while requiring many layers. The specific design makes it a reasonable choice based on the capacitance value size and the filter space layout.

Table 2. Comparison of two structures with the same effective capacitance value.

Structural Form	MIM Capacitor	VIC Capacitor
Occupied area	Large	Small
Self-resonant frequency	Slightly low	High
Q value	Slightly lower	High
Number of layers required	less	Multi

Regardless of the capacitor design framework, the relevant parameters of the capacitor element should be extracted. The capacitor is considered a two-port network, and its corresponding characteristic parameters are extracted, yielding a precise construction of the capacitor framework and dimensions. Since the capacitors in the filter have small capacitance values, i.e., C2 = 0.139 pF and C3 = 0.506 pF, the embedded MIM framework is utilized. Considering C1 = 7.163 pF, a relatively large capacitance, an embedded VIC framework is utilized to decrease the planar area.

After calculating the conductance parameter Y by HFSS, the capacitance value and quality factor parameters can be obtained as follows:

$$C_{eff} = -\frac{Im(Y(2,1))}{2\pi \cdot freq} \tag{16}$$

$$Q = -\frac{Im(Y(1,1))}{Re(Y(1,1))} \tag{17}$$

where C_{eff} is the effective value of the capacitor, and freq is the self-resonant frequency.

HFSS simulation determines the physical dimensions of C1, C2, and C3. The 3D structure of C1 is presented in Figure 11a, which has orange and green colours to distinguish the two polarities of the capacitor and the capacitance simulation results are presented in Figure 11b.

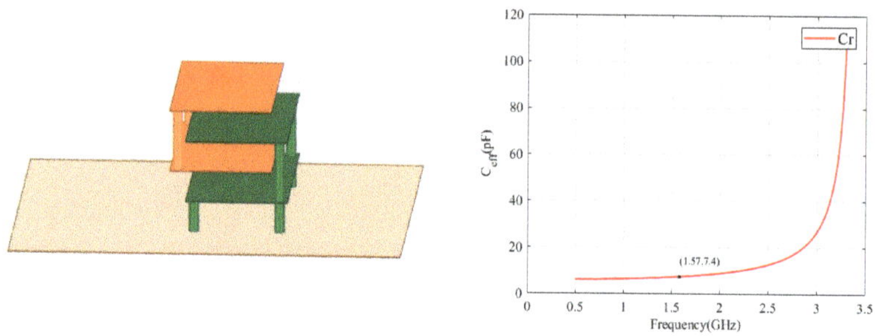

(**a**) (**b**)

Figure 11. 3D structure diagram (**a**) and simulation results of capacitor C1 (**b**).

3.2. Inductors

LTCC inductors generally include planar and spiral types. Since the Q value of the inductor is much lower than the capacitance, the inductor design plays an essential role in the filter insertion loss.

Table 3 compares the main performance and parameters of the four inductor structures. It can be seen that with the same effective inductance value, the spiral inductor has the advantages of a relatively higher resonant frequency and quality factor, a smaller single-layer area, and the disadvantage of the number of required layers, followed by the stacked, displacement, and planar structures. The specific design should choose the inductor implementation form according to the key parameters. The inductor effective value (Leff) determines the size of the total inductance, and the Q value determines the size of the inductor loss. The design of LTCC inductors should focus on the effective value of the inductance, quality factor, and self-resonant frequency SRF; the effective value of the inductance determines the size of the inductor and the specific application of the resonance frequency, the quality factor determines the loss of RF devices and other performance parameters, and the self-resonant frequency SRF determines the effective use of the inductor frequency range.

Table 3. Comparison of inductance performance of various structures with the same inductance value.

Type	Structural Form	Occupied Area	Self-Resonant Frequency	Q Value	Number of Floors
Monolayer	Planar	Maximum	Lowest	Lowest	least
	Displacement	Small	High	High	Medium
Multilayer	Stacked	Medium	Medium	Medium	Medium
	Spiral	Smallest	Highest	Highest	Most

The inductance values of L1 and L2 in the filter model are small and can be realized by a single layer of short wires or through holes to avoid the complexity caused by mutual coupling with other capacitors. The inductor's width is 0.2 mm, i.e., the 50 Ω stripline's width. HFSS simulation is employed to determine the inductor's length. Inductor L1 is the main inductor in the filter and directly affects the center frequency. A planar-type inductor is utilized to prevent mutual coupling with the other capacitors.

The LTCC's effective inductance can be described in terms of admittance parameters as follows [16]:

$$L_{eff} = \frac{Im\left(\frac{1}{Y(1,1)}\right)}{2\pi \cdot freq} \tag{18}$$

$$Q = -\frac{Im(Y(1,1))}{Re(Y(1,1))} \tag{19}$$

where freq denotes the self-resonant frequency, and Y11 is the inductor's conductance value, calculated by HFSS. Figure 12 presents the inductor's 3D model, with L1 = 1.82 nH. The conductor's width in the inductor is 0.2 mm, while the total length is 2.6 mm. Figure 12 presents the effective inductance attained through HFSS simulation.

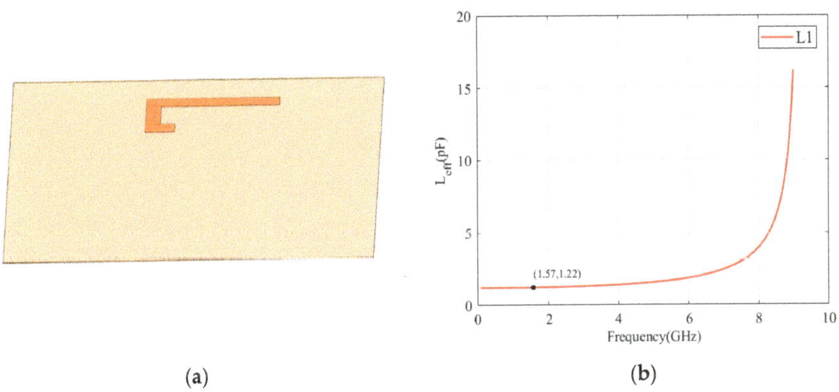

Figure 12. 3D structure diagram (**a**) and simulation results of inductor L1 (**b**).

3.3. Simulation Results

Based on the inductor and capacitor models, all of the filter components are modeled, simulated, and optimized in 3D. The three-dimensional structure simulation is as close as possible to the actual situation; a FerroA6M material is selected as the substrate material, the metal line material is set to gold, and the thickness is 8 μm. As shown in Figure 7, there are four inductors and five capacitors with different capacitance values. These nine variables pose a great difficulty in the modeling and optimization process. In order to facilitate modeling and optimization, the current study employs a symmetric circuit framework. Nine variables are reduced to five using the model's symmetry, thus simplifying the LTCC. Figure 13 presents the LTCC bandpass filter's 3D model. It is worth mentioning that to show the picture more clearly, the model layer spacing is enlarged five times, and the other dimensions and the layer spacing are kept unchanged. The 3D model parameters are as follows:

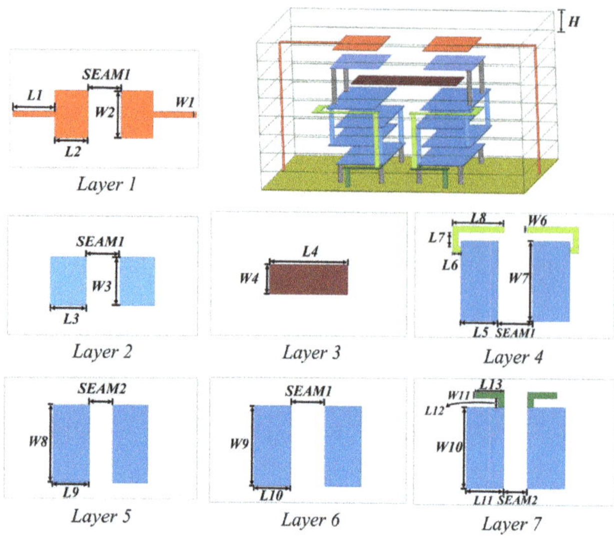

Figure 13. 3D modeling of LTCC bandpass filters.

SEAM = L2 = 1.2 mm, L1 = 1.5 mm, W1 = W6 = W11 = 0.2 mm, W2 = W3 = 1.7 mm, L4 = 3.6 mm, W4 = L13 = 1 mm, L6 = L12 = 0.3 mm, L7 = 0.5 mm, L8 = 1.8 mm, W7 = W8 = W9 = W10 = 2.7 mm, L3 = L5 = L9 = L10 = L11 = 1.3 mm, and H = 97 μm.

By verifying the simulation outcomes of the flexible layout of inductors and capacitors in the model, the model is continuously optimized and adjusted to obtain the optimum simulation results, as presented in Figure 14. The simulation results indicate that in the 1.525–1.625 bandwidth, the insertion loss is relatively slight, the return loss meets the requirements, and the TZ is near 6 GHz. However, since there is a resonance near 9.5 GHz, a new bandstop filter should be introduced to eliminate the resonance.

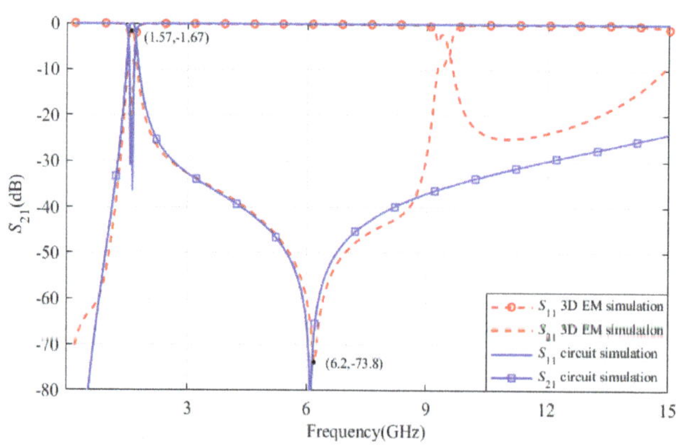

Figure 14. LTCC bandpass filter simulation results.

4. Wide Stopband Filter Design

4.1. Principle of the Wide Stopband Filter

The metrics in Table 1 indicate the relatively high out-of-band rejection requirement of the filter. There are two methods to improve the filter's out-of-band rejection when designing the filter: increasing the filter's order and adding the TZs. However, as the order increases, the filter size increases [17], which is inappropriate for designing miniaturized filters. Therefore, transmission zeros are generally added when constructing the miniaturized filters [18,19]. The filter's zero frequency is the frequency that makes the transmission equation zero. Ideally, since energy cannot pass through the zero frequency, excellent isolation can be attained. The set parameter mode generally adopts series or parallel resonances.

When the capacitance and inductance in a circuit are connected in series or parallel, the impedance and conductance in a high-frequency circuit are calculated as follows:

$$jX(\omega) = j\omega C + \frac{1}{j\omega L} \tag{20}$$

$$jB(\omega) = j\omega L + \frac{1}{j\omega C} \tag{21}$$

According to Equation (20), an open circuit is obtained as the conductance in a parallel resonant circuit is zero, entirely reflecting the energy at that frequency, thus constructing a transmission zero. According to Equation (21), when the LC satisfies the resonance relation, the input impedance in the series resonant circuit becomes zero, causing a short circuit. The whole energy at the mentioned frequency is absorbed, forming a TZ.

4.2. Three-Dimensional Structural Design

This bandstop filter adopts the form of Figure 15b; the inductor is realized using a narrow wire inductor with an over-hole, and the capacitor is only realized using a flat-plate capacitor. Figure 16 shows the three-dimensional structural model diagram.

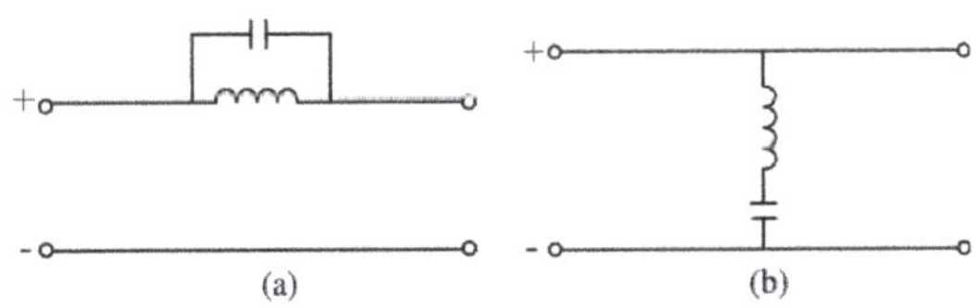

Figure 15. (**a**) Parallel resonant circuit. (**b**) Series resonant circuit.

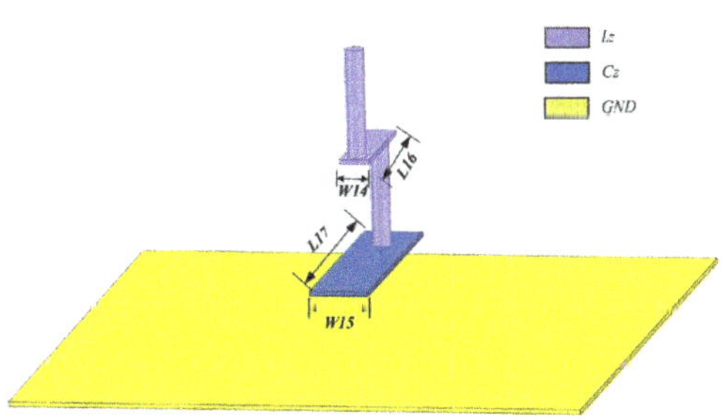

Figure 16. 3D model diagram of the band reject filter.

4.3. Simulation Results

The band reject filter can change the resonance point by adjusting the inductor length Lz or the flat size of the capacitor Cz, as shown in Figure 16, by changing the capacitor width W15 and thus changing the resonance frequency point. As W15 increases, the capacitance Cz increases, and the resonance point is shifted to a lower frequency, compatible with the theoretical analysis. Figure 17 shows the simulation results. This simulation is only performed for the bandstop filter. The following section presents the adjustment of the transmission zeros of the target filter after cascading the narrowband filter with the bandstop filter.

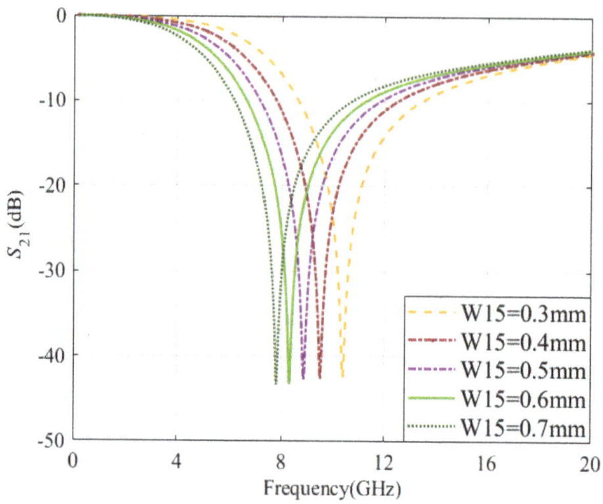

Figure 17. Simulation results of the band reject filter.

5. Overall Simulation

5.1. Simulation Results

The model diagram after cascading the bandpass filter designed in Section 3 with two new bandstop filters, as shown in Figure 18, is designed to achieve the wide stopband index requirement. Figure 18 shows the equivalent circuit diagrams of the corresponding lumped elements. Figure 19 marks different lumped elements with different colors, each corresponding to that in Figure 18, and the overall structure shows symmetry. The model parameters are W12 = W14 = 0.2 mm, L14 = L16 = 1 mm, L15 = 2 mm, W13 = W15 = 0.5 mm, and L17 = 1.6 mm. Figure 20 shows the exterior view of the filter when placed on an RO4350B substrate. RO4350B can be well applied to high-frequency circuits. The substrate thickness is h = 0.508, and the dielectric constant is ε = 3.66. The transmission line to the filter inputs and outputs is of the CPW ground-type, matched to 50 ohms to match the filter inputs and outputs. For better matching, tapered transmission lines are added close to the input and output of the filter. The line width is 0.48 mm, and the line gap is 0.24 mm. As shown in Figure 21, the transmission zeros of the target filter are adjusted by adjusting the size of the capacitor Cz. The results indicate that as L15 decreases, the transmission zeros move to higher frequencies. When the frequency of transmission zero 3 (TZ3) is too high, the transmission zeros should be placed appropriately to successfully suppress the harmonics. The second transmission zero 2 (TZ2) is adjusted similarly to TZ3 by controlling TZ2 and adjusting the size of the capacitance on the other side. Figure 10 illustrates the control of the first transmission zero 1 (TZ1). The filter with the best stopband characteristics is obtained by adjusting the three transmission zeros. Figure 22 shows the HFSS simulation of this bandpass filter. As shown in Figure 22, compared with ADS simulation, these two simulations have relatively close transmission results, indirectly demonstrating the feasibility of the equivalent circuit.

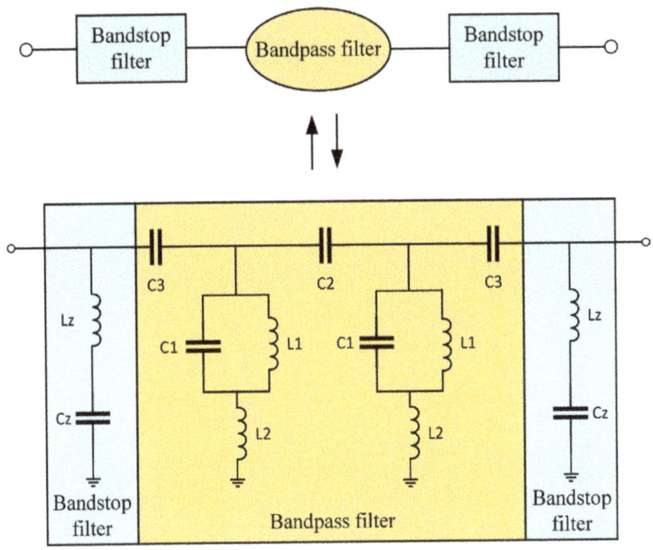

Figure 18. Wide stopband narrow bandpass filter schematic diagram.

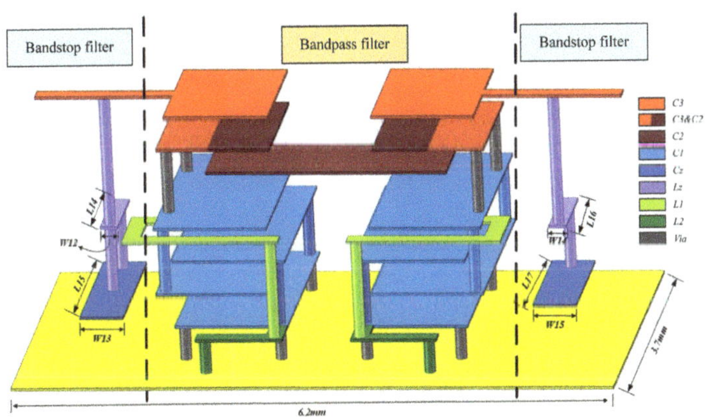

Figure 19. Three-dimensional structure of a wide stopband narrow bandpass filter.

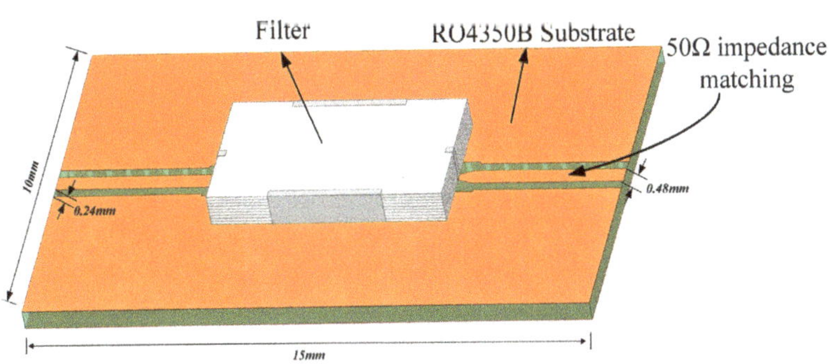

Figure 20. Exterior view of the wide stopband narrow bandpass filter.

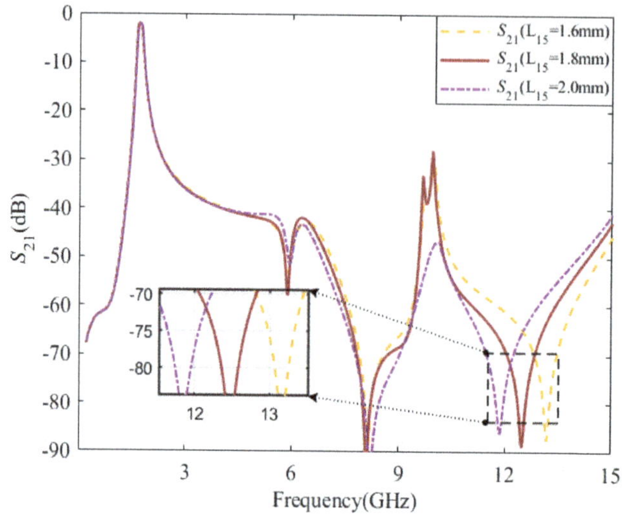

Figure 21. Effect of L15 on the filter's high-frequency transmission zero 3.

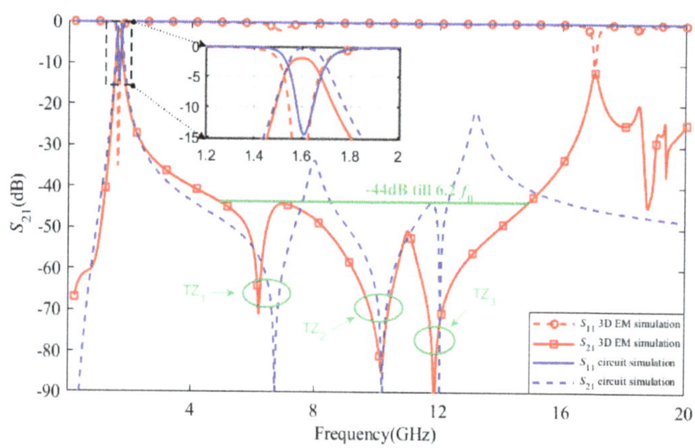

Figure 22. Simulation results of the wide stopband narrow bandpass filter.

5.2. Performance Comparison

Table 4 compares the efficiency of the constructed L-band bandpass filter with previously published studies.

Table 4. Comparison with previous studies.

Refs.	F0 (GHz)	FBW (%)	IL (dB)	Stopband Bandwidth	Rejection Level (dB)	Core Size (λ_g^2)	Selectivity Factor (%)
[20]	0.835	15.8%	0.41	$4.26f_0$	<20 dB	0.16 ∗ 0.12	32.51@20 dB
[21]	5.5	18.2%	2.46	$1.02f_0$	<32 dB	0.058 ∗ 0.029	9.9@32 dB
[22]	0.75	12.8%	0.6	$5.2f_0$	<20 dB	0.014 ∗ 0.02	5.87@20 dB
[23]	6.11	8.96	1.07	$2.05f_0$	<22 dB	0.153 ∗ 0.194	32.22@20 dB
[24]	3.5	22.9%	0.5	$1.64f_0$	<18 dB	0.09 ∗ 0.09	13.0@20 dB
This work	1.57	6.3%	1.65	$6.2f_0$	<44 dB	0.036 ∗ 0.022	110@20 dB 75.98@40 dB

Table 4 shows that compared with the studies performed in the last five years (the studies [20,21,23,24]), the proposed filter occupies less space by stacking the collector elements in the Z-direction through the LTCC process. Compared with the literature [20–24], focusing on wide stopband bandpass filters, the designed filter provides better stopband characteristics and narrower bandwidth, with moderate insertion loss. As shown in Table 4, the proposed filter has an excellent selectivity factor (SF), which is crucial for wide stopband filters.

6. Conclusions

The current study describes a bandpass filter with narrowband and considerable out-of-band rejection for the GPS band. The filter framework can be simplified and the simulation complexity can be alleviated using symmetry. By the new filter prototype and cascading the new three-dimensional structure of the bandstop filter, three TZs are produced to promote the filter's stopband characteristics significantly. LTCC technology has a significant advantage in realizing the filter's miniaturization, leading to a size of 6.2 mm $*$ 3.7 mm $*$ 0.78 mm, smaller than the conventional process filter. The whole circuit has the advantages of high performance, small size, and low cost (in the case of mass production), making it suitable for GPS wireless communication-integrated systems and GPS L1 frequency pre-filtering applications.

In the future development of integration technology, the requirements for the performance and miniaturization of communication systems will also increase. Thus, creating miniaturized filters with LTCC technology will be one of the future development trends. In addition, with the development of wireless technology, the spectrum's utilization is becoming increasingly important. Therefore, filtering signals at other frequencies will also become more and more critical.

Author Contributions: Conceptualization, Y.M., Q.D., W.Z. and C.L.; methodology, Y.M. and Q.D.; software, Y.M.; validation, Y.M., W.Z., C.L. and H.Z.; investigation, Y.M., Q.D. and C.L.; writing—original draft preparation, Y.M. and Q.D.; writing—review and editing, Y.M., Q.D., W.Z., C.L. and H.Z. All authors have read and agreed to the published version of the manuscript.

Funding: This research received no external funding.

Data Availability Statement: https://www.scidb.cn/en/s/2MFZFb (accessed on 8 November 2023).

Acknowledgments: Qifei Du and Cheng Liu are gratefully acknowledged for their insightful comments and helpful discussions.

Conflicts of Interest: The authors declare no conflicts of interest.

References

1. Ihle, M.; Ziesche, S.; Zech, C.; Baumann, B. Compact LTCC packaging and printing technologies for Sub–THz modules. In Proceedings of the 2018 7th Electronic System-Integration Technology Conference (ESTC), Dresden, Germany, 18–21 September 2018.
2. Tang, J.M.; Liu, H.W.; Yang, Y. Balanced dual-band superconducting filter using stepped-impedance resonators with high band-to-band isolation and wide stopband. *IEEE Trans. Circuits Syst. II Express Briefs* **2021**, *68*, 131–135. [CrossRef]
3. Xia, L.; Wu, B.; Song, L.; Chen, J.Z.; Cheng, Q.S. Design of pseudoelliptic filters with controllable transmission zeros using high-Q double-layer suspended stripline resonators. *Int. J. RF Microw. Comput. Aided Eng.* **2019**, *29*, e21785. [CrossRef]
4. Xu, K.D.; Xu, H.; Liu, Y.H.; Ai, J.; Liu, Q.H. Short- and open-stub loaded spiral resonator and its application in planar microstrip filters. *IET Microw. Antennas Propag.* **2017**, *11*, 363–369. [CrossRef]
5. Kim, C.H.; Chang, K. Wide-stopband bandpass filters using asymmetric stepped-impedance resonators. *IEEE Microw. Wirel. Compon. Lett.* **2013**, *23*, 69–71. [CrossRef]
6. Arabi, E.; Lahti, M.; Vähä-Heikkilä, T.; Shamim, A. A 3-D miniaturized high selectivity bandpass filter in LTCC technology. *IEEE Microw. Wirel. Compon. Lett.* **2014**, *24*, 8–10. [CrossRef]
7. Tan, C.; Yu, Z.J.; Xie, C.S. An L-band bandpass filter with narrow bandwidth and miniature based on LTCC technology. *Prog. Electromagn. Res.* **2019**, *87*, 63–72. [CrossRef]
8. Peng, Z.; Zhou, J.; Rao, Y.B.; Deng, Z.X.; Luo, X. Compact bandpass filter with wide stopband and low radiation using fully-packaged substrate integrated defected ground structure. In Proceedings of the 2021 IEEE International Symposium on Radio-Frequency Integration Technology (RFIT), Hualien, Taiwan, 25–27 August 2021.

9. Li, S.X.; Ma, K.X.; Chen, X.; Yan, N.N.; Fu, H.P. A miniaturized bandpass filter with wideband and high stopband rejection using LTCC technology. In Proceedings of the 2020 IEEE Asia-Pacific Microwave Conference (APMC), Hong Kong, China, 8–11 December 2020.
10. Feng, W.J.; Gao, X.; Che, W.Q.; Yang, W.C.; Xue, Q. LTCC wideband bandpass filters with high performance using coupled lines with open/shorted stubs. *IEEE Trans. Compon. Packaging Manuf. Technol.* **2017**, *7*, 602–609. [CrossRef]
11. Dai, Y.S.; Xu, L.; Han, Q.F.; Chen, S.B.; Wang, L.J.; Chen, R.Q. Miniaturized LTCC wideband bandpass filter using lumped-element shunt LC resonators. In Proceedings of the 2012 International Conference on Microwave and Millimeter Wave Technology (ICMMT), Shenzhen, China, 5–8 May 2012.
12. Du, J.C.; Chen, N.; Lu, Q.Y. Design of a Ka-band Microstrip quasi-Yagi antenna array based on LTCC substrate. In Proceedings of the 2013 IEEE International Conference on Microwave Technology & Computational Electromagnetics, Qingdao, China, 25–28 August 2013.
13. Matthaei, G.L.; Young, L.; Jones, E.M.T. *Microwave Filters, Impedance-Matching Networks, and Coupling Structures*; Artech House, Inc.: Boston, MA, USA, 1980.
14. Hong, J.S.; Lancaster, M.J. Advanced RF/microwave filters. In *Microstrip Filters for RF/Microwave Applications*; Hong, J.S., Lancaster, M.J., Eds.; John Wiley & Sons, Ltd.: New York, NY, USA, 2001; pp. 315–377.
15. Leung, W.Y.; Cheng, K.K.M.; Wu, K.L. Multilayer LTCC bandpass filter design with enhanced stopband characteristics. *IEEE Microw. Wirel. Compon. Lett.* **2002**, *12*, 240–242. [CrossRef]
16. Wu, J.J.; Yan, B.; Zhang, Q.S.; Guo, G.F.; Su, H. Design and parameters extraction methods of microwave LTCC embedded inductor. *Telecommun. Eng.* **2007**, 123–126. (In Chinese)
17. Zhu, L.F. Narrowband LTCC filter with length-reduced end-coupled resonators. *Prog. Electromagn. Res. Lett.* **2020**, *93*, 13–19. [CrossRef]
18. Dai, Y.S.; Zhang, H.S.; Qi, G.P.; Lu, D.L.; Tao, Z.L.; Qian, G.X.; Wang, H. LTCC bandpass filter for Bluetooth^{TM} application with dual transmission zeros. In Proceedings of the 2008 International Conference on Microwave and Millimeter Wave Technology, Nanjing, China, 21–24 April 2008.
19. Cui, L.W.; Yu, J.G.; Li, S.L.; Liu, Y.N. A sharp-rejection dual-bandstop filter with multiple transmission poles and zeros based on transversal signal-interaction concepts. In Proceedings of the 2017 IEEE 5th International Symposium on Electromagnetic Compatibility (EMC-Beijing), Beijing, China, 28–31 October 2017.
20. Barik, R.K.; Koziel, S.; Pietrenko-Dabrowska, A. Broad stopband, low-loss, and ultra-compact dual-mode bandpass filter based on HMSIRC. *Electronics* **2023**, *12*, 2831. [CrossRef]
21. Li, Y.L.; Zhu, X.; Liu, J.C.; Zhou, L.J.; Wang, Z.H. Miniaturization of low temperature co-fired ceramic packaging for microwave filters. In Proceedings of the 2018 19th International Conference on Electronic Packaging Technology (ICEPT), Shanghai, China, 8–11 August 2018.
22. Chen, L.; Yang, Z.H.; Ling, S.X.; Ding, D.W. Design of LTCC bandpass filter with broadband external suppression for 5G applications. In Proceedings of the 2023 International Conference on Microwave and Millimeter Wave Technology (ICMMT), Qingdao, China, 14–17 May 2023.
23. Duraisamy, T.; Barik, R.K.; Cheng, Q.S.; Kamakshy, S.; Sholampettai Subramanian, K. Miniaturized SIW filter using D-shaped resonators with wide out-of-band rejection for 5G applications. *J. Electromagn. Waves Appl.* **2020**, *34*, 2397–2409. [CrossRef]
24. Tie, H.Y.; Luo, L.; Song, Z.C.; Ma, Q.; Hu, S.W.; Zhou, B. Dual-mode LTCC filter with 4th-order harmonic suppression for 5G N78 applications. In Proceedings of the 2021 International Conference on Microwave and Millimeter Wave Technology (ICMMT), Nanjing, China, 23–26 May 2021.

electronics

MDPI

Article

Analysis and Optimization Design Scheme of CMOS Ultra-Wideband Reconfigurable Polyphase Filters on Mismatch and Voltage Loss

Yingze Wang [1,†], Xiaoran Li [1,2,3,4,†], Yuanze Wang [1], Xinghua Wang [1,2,3,4,*], Zicheng Liu [3], Fang Han [4] and Quanwen Qi [4]

1 School of Integrated Circuits and Electronics, Beijing Institute of Technology, Beijing 100081, China; 3120210708@bit.edu.cn (Y.W.); xiaoran.li@bit.edu.cn (X.L.); 3220221603@bit.edu.cn (Y.W.)
2 Advanced Technology Research Institute, Beijing Institute of Technology, Jinan 250101, China
3 BIT Chongqing Institute of Microelectronics and Microsystems, Chongqing 401332, China; liuzc@bit.edu.cn
4 Yangtze Delta Region Academy of Beijing Institute of Technology, Jiaxing 314000, China; hanfang@bit.edu.cn (F.H.); 7520200244@bit.edu.cn (Q.Q.)
* Correspondence: 89811@bit.edu.cn
† These authors contributed equally to this work.

Abstract: This manuscript presents an analysis and optimization scheme for the ultra-wideband passive reconfigurable polyphase filters (PPFs) to minimize I/Q (in-phase and quadrature-phase) phase/amplitude mismatch and voltage loss. By building a mathematical model of the voltage transfer, the relationship between the resonant frequency of each stage and the I/Q mismatch and the relationship between the network impedance and the voltage loss are revealed, providing a scheme for PPF optimization. The proof-of-concept 2~8 GHz wideband reconfigurable PPF is designed in a 55 nm CMOS process. The optimization scheme enables the designed PPF to achieve an I/Q phase mismatch within 0.2439° and an I/Q amplitude mismatch within 0.098 dB throughout the entire band, and it shows great robustness during Monte Carlo sampling. The maximum voltage loss is 17.7 dB, and the total chip area is 0.174×0.145 mm^2.

Keywords: PPF; I/Q phase/amplitude mismatch model; voltage loss model; reconfiguration

1. Introduction

With the development of communication technology, transceivers with quadrature mixers are being used more and more widely. As all such transceivers need to provide orthogonal local carriers, the design of the in-phase and quadrature-phase signal (I/Q) generator is very important in the transceiver. Figure 1 illustrates the fundamental schematic for the local oscillator (LO). The single-phase signal is initially directed to a single-to-differential circuit to obtain differential signals. Subsequently, a quadrature generator is employed to derive I/Q signals, followed by signal amplification through a buffer.

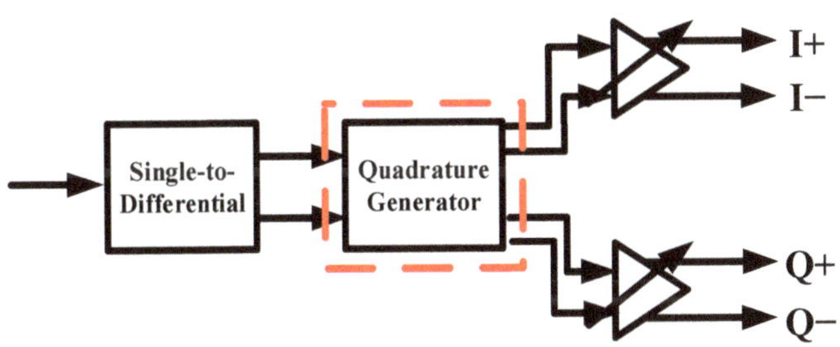

Figure 1. Application of quadrature signal generator in LO generation circuit.

There are four main ways to generate RF quadrature signals: the RC-CR network, frequency dividers consisting of master–slave flip flops, orthogonal LC oscillators, and the polyphase network. Compared with the other three, the polyphase network is relatively insensitive to the mismatch of components and has a wider bandwidth compared to the RC-CR circuit [1]. At the same time, without an inductor in the circuit, the area is small. Moreover, it has no frequency division, which simplifies the design of the signal generator [2]. Therefore, the polyphase network has been widely used to generate I/Q signals [3–6]. As one of the polyphase networks, the PPF has attracted the attention of many researchers due to its advantages in jitter and power performance. Currently, it is commonly used for applications in Ka and lower bands.

However, there are two challenges to be solved in the PPF design: how to reduce phase and amplitude mismatch and how to reduce voltage loss. Reference [7] proposes an intermediate point compensation method to improve I/Q quadrature; it detects the variation and compensates the quadrature phase mismatch by feeding the network's output to the intermediate point. In [8], it shorts half of the capacitors and resistors of the first stage to reduce the loss and area. In [9], it adds two inductors at the output of the PPF circuit to reduce the loss, at the cost of increasing the occupied area. However, there are few papers that discuss the performance tuning based on the PPF's own parameters. Reference [10] delivers a mathematical model for passive PPF design; however, it lacks the cooperation with practical applications, including layout and post-simulations.

To improve the performance of the PPF from the perspective of the basic design process, based on the mathematical modeling, this paper analyses the model of the voltage transfer function of the PPF output signals at length and introduces the impact of the load. Also, some comparisons are drawn between the model in this paper and that in [10]. After that, a general design and optimization scheme for the PPF is given, including ultra-wideband PPF reconfiguration, resonant frequency decision, and voltage loss reduction, etc. And to verify the feasibility of the scheme, this paper presents a design of a reconfigurable PPF working at 2~8 GHz.

The structure of this paper is as follows. In Section 2, the basic model of the PPF is established, and the general rule of the PPF orthogonal phase error is analyzed in detail. In Section 3, the input and output impedance model of the PPF is proposed and is utilized to analyze the loss of the multistage PPF. Section 4 shows the PPF design process of 2~8 GHz broadband and gives the general process of the PPF design. Section 5 gives the results of the post-layout simulation. Finally, the conclusion is given in Section 6.

2. PPF I/Q Mismatch Model

2.1. Basic Principle of Multistage PPF

Two typical single-stage PPF structures are shown in Figure 2, one with two inputs connected together (PPF-I) and the other with two inputs connected to ground (PPF-II).

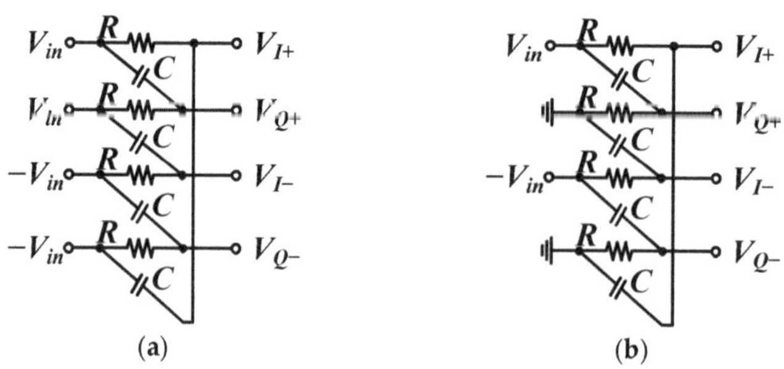

Figure 2. Single-stage PPF: (**a**) PPF-I: two inputs connected together; (**b**) PPF-II: two inputs connected to ground.

Traditionally, for the PPF shown in Figure 2a, the equations of the four outputs are those shown in Equations (1)–(4).

$$V_{I+} = \frac{(-CR)\omega - \mathrm{i}}{(CR)\omega - \mathrm{i}} V_{in} \tag{1}$$

$$V_{Q+} = V_{in} \tag{2}$$

$$V_{I+} = \frac{(CR)\omega + \mathrm{i}}{(CR)\omega - \mathrm{i}} V_{in} \tag{3}$$

$$V_{Q-} = -V_{in} \tag{4}$$

Then, according to Equations (1)–(4), the output ratio of V_I and V_Q is

$$\frac{V_I}{V_Q} = \frac{(-CR)\omega - \mathrm{i}}{(CR)\omega - \mathrm{i}} = -\frac{\omega + \mathrm{i}\omega_0}{\omega - \mathrm{i}\omega_0} \tag{5}$$

where $\omega_0 = 1/RC$.

Similarly, for the PPF shown in Figure 2b, the output ratio of V_I and V_Q is as shown in Equation (6).

$$\frac{V_I}{V_Q} = \frac{\omega}{\omega_0}\mathrm{i} \tag{6}$$

From Equations (5) and (6), it is evident that the I/Q output signals in PPF-I consistently maintain their amplitude across all the frequencies but exclusively exhibit quadrature phases at the resonant frequency (1/RC). Conversely, the I/Q output signals in PPF-II retain quadrature phases at all the frequencies but share the same amplitude solely at the resonant frequency (1/RC).

Notably, although the image rejection ratio (IRR) is the commonly preferred parameter in typical measurements, the analysis in [2] underscored the PPF's IRR dependence on the quadrature phase and amplitude mismatches in the output signals. Consequently, an intuitive observation of quadrature phase/amplitude mismatches in the output signal is presented in the analysis in the following sections instead of the IRR.

Next, advancing to the analysis of the multistage PPF, the voltage ratio between V_I and V_Q is determined by multiplying the matrix chains of each stage. Figure 3 illustrates the consistent multistage PPF matrix for both PPF-I and PPF-II.

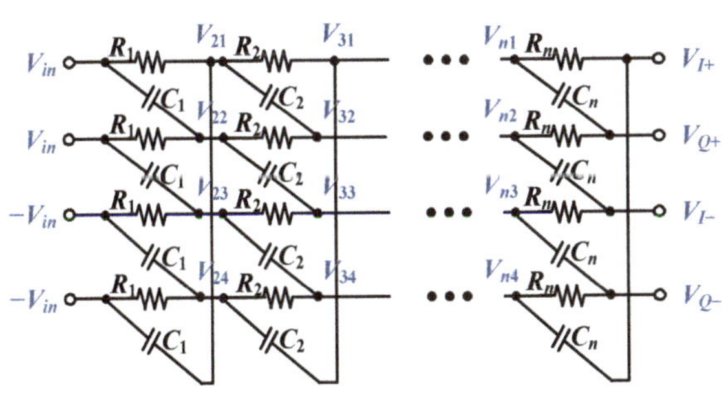

Figure 3. Matrix of multistage PPF.

According to Figure 3, the recursive matrix can be obtained, as shown in Equation (7).

$$\begin{aligned}\begin{bmatrix} V_{n+1,1} \\ V_{n+1,2} \\ V_{n+1,3} \\ V_{n+1,4} \end{bmatrix} &= \begin{bmatrix} 1 & 0 & 0 & iC_nR_n\omega \\ iC_nR_n\omega & 1 & 0 & 0 \\ 0 & iC_nR_n\omega & 1 & 0 \\ 0 & 0 & iC_nR_n\omega & 1 \end{bmatrix} \begin{bmatrix} V_{n,1} \\ V_{n,2} \\ V_{n,3} \\ V_{n,4} \end{bmatrix} \frac{1}{1+iC_nR_n\omega} \\ &= \begin{bmatrix} \omega_n & 0 & 0 & i\omega \\ i\omega & \omega_n & 0 & 0 \\ 0 & i\omega & \omega_n & 0 \\ 0 & 0 & i\omega & \omega_n \end{bmatrix} \begin{bmatrix} V_{n,1} \\ V_{n,2} \\ V_{n,3} \\ V_{n,4} \end{bmatrix} \frac{1}{\omega_n+i\omega}\end{aligned} \tag{7}$$

where $\omega_n = 1/R_nC_n$.

The only difference is that the calculation of $V_{n,1\ldots4}$ for PPF-I and PPF-II is $V_{1,1\ldots4}$, which is $[V_{in+}\ V_{in+}\ V_{in-}\ V_{in-}]$ for the former and $[V_{in+}\ 0\ V_{in-}\ 0]$ for the latter.

Now, using Equation (7), the output signals of each stage can be calculated in both PPF-I and PPF-II. Next, the three-stage PPF is calculated and discussed based on Equation (7). However, any specific calculation process will no longer be presented in detail to ensure the succinctness of this article.

For the PPF shown in Figure 4a, the output ratio of V_I and V_Q is shown in Equation (8).

$$\frac{V_I}{V_Q} = \frac{-i\omega^3 + (\omega_1+\omega_2+\omega_3)\omega^2 - (\omega_1\omega_2+\omega_1\omega_3+\omega_2\omega_3)i\omega + \omega_1\omega_2\omega_3}{i\omega^3 + (\omega_1+\omega_2+\omega_3)\omega^2 + (\omega_1\omega_2+\omega_1\omega_3+\omega_2\omega_3)i\omega + \omega_1\omega_2\omega_3} \tag{8}$$

where $\omega_1 = 1/R_1C_1$, $\omega_2 = 1/R_2C_2$, $\omega_3 = 1/R_3C_3$.

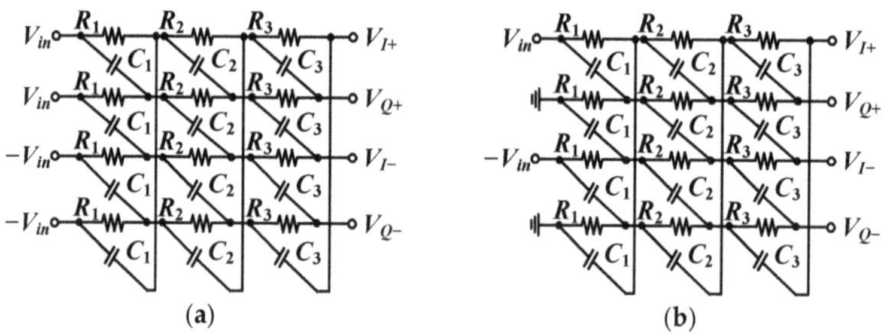

Figure 4. Three-stage PPF: (**a**) PPF-I: two inputs connected together; (**b**) PPF-II: two inputs connected to ground.

Similarly, for the PPF shown in Figure 4b, the output ratio of V_I and V_Q is shown in Equation (9).

$$\frac{V_I}{V_Q} = \frac{(\omega_1+\omega_2+\omega_3)\,\omega^2 + \omega_1\,\omega_2\,\omega_3}{i\,\omega^3 + (\omega_1\omega_2\,+\omega_1\omega_3\,+\omega_2\omega_3\,)i\,\omega} \tag{9}$$

Based on Equations (8) and (9), the phase and magnitude mismatches of the I/Q signals of two kinds of PPF are plotted as shown in Figure 5.

From Figure 5a and Equation (8), it can be seen that for PPF-I, the I/Q output signals have the same amplitude at all frequencies and have quadrature phases at three frequency points, which correspond to the resonant frequency ($1/RC$) of each stage. Similarly, from Figure 5b and Equation (9), for PPF-II, the I/Q output signals have quadrature phases at all frequencies and have the same magnitude at three frequency points, which also correspond to the resonant frequency ($1/RC$) of each stage.

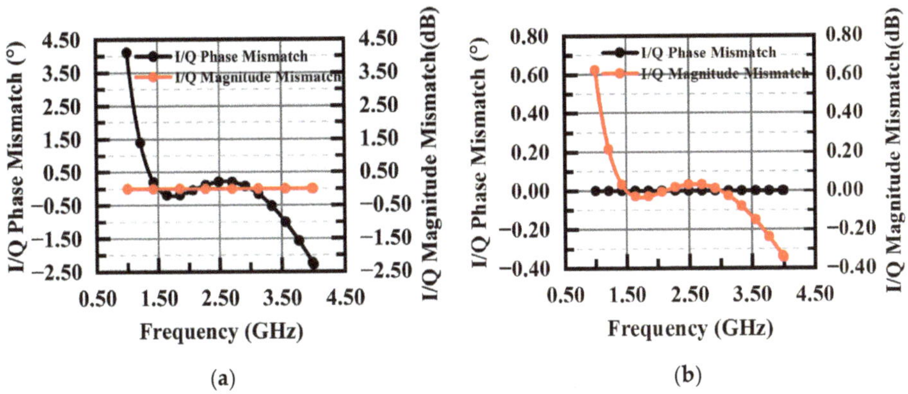

Figure 5. I/Q mismatch of output signals in (**a**) PPF-I and (**b**) PPF-II.

From the analysis above, neither PPF-I nor PPF-II can achieve full-band orthogonality, but they can be optimized through proper design; this is shown in the next section.

2.2. Design Scheme for I/Q Mismatch Reduction in Multistage PPF

According to Equations (8) and (9), to reduce the I/Q mismatch, resonant frequencies should play a crucial role in the multistage PPF.

Take the structure of PPF-I as an example. In order to find the relationship between the resonant frequencies and the phase mismatch, a simulation is carried out based on Equation (8), where the intermediate frequency point ($f_2 = \omega_2/2\pi$) is swept and the corresponding phase mismatch is obtained, under the condition that the low frequency point ($f_1 = \omega_1/2\pi$) and the high frequency point ($f_3 = \omega_3/2\pi$) are fixed. Note that since the magnitude mismatch theoretically remains the same for PPF-I, there is no need to analyze the magnitude mismatch in the mathematical model.

The sweeping result is shown in Figure 6a, where the pass band is also changed by increasing f_3 from 5 GHz to 10 GHz to find the universality pattern. The abscissa is the frequency of f_2 (in GHz), and the ordinate is the maximum I/Q phase mismatch throughout f_1~f_3 (in °). It can be seen that when the minimum full-band phase mismatch is achieved, f_2 is always $\sqrt{f_1 f_3}$. The fitting function is shown in Figure 6b.

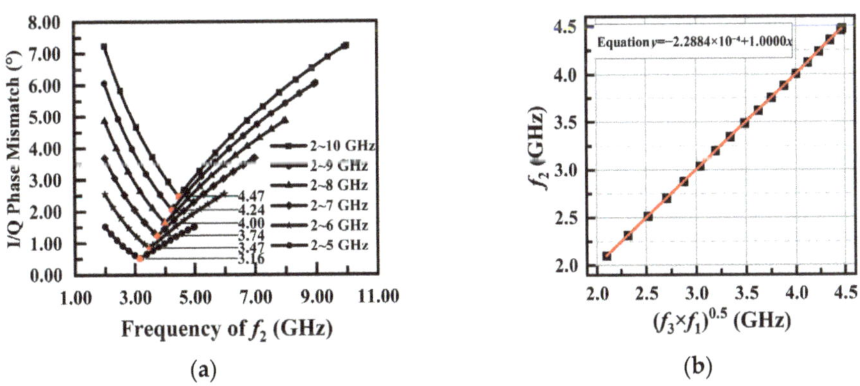

Figure 6. I/Q mismatch of output signals for PPF-I: (**a**) relationship between f_2 and maximum I/Q phase mismatch throughout f_1~f_3; (**b**) fitting function between f_2 and $\sqrt{f_1 f_3}$.

Similarly, for the structure of PPF-II, the intermediate frequency point ($f_2 = \omega_2/2\pi$) is also swept to obtain the corresponding magnitude mismatch; the result is shown in Figure 7. The fitting function is almost the same as that of PPF-I, which indicates that $f_2 \approx \sqrt{f_1 f_3}$.

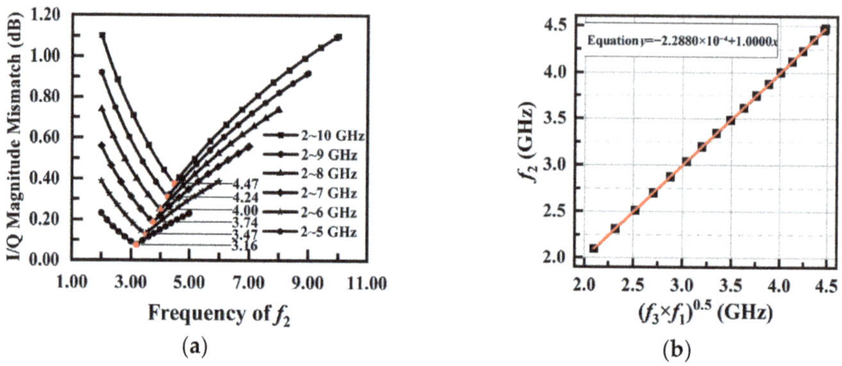

Figure 7. I/Q mismatch of output signals for PPF-II: (**a**) relationship between f_2 and maximum I/Q magnitude mismatch throughout f_1~f_3; (**b**) fitting function between f_2 and $\sqrt{f_1 f_3}$.

Now, the conclusion is that in order to achieve good quadrature performance, whether in the structure of PPF-I or PPF-II, the intermediate frequency point ω_2 should be

$$\omega_2 = \sqrt{\omega_1 \omega_3} \tag{10}$$

Analogously, for the PPF with more than three stages, to reach the lowest I/Q mismatch, the relationship between the resonant frequencies should be

$$\omega_2 = \sqrt{\omega_1 \omega_3},\ \omega_3 = \sqrt{\omega_2 \omega_4},\ \omega_4 = \sqrt{\omega_3 \omega_5} \cdots \tag{11}$$

It should be noted that since $\omega = 1/(RC)$, the phase mismatch is only related to the product of each resistance and capacitance, not their individual values.

2.3. Design Scheme for I/Q Mismatch Reduction in Reconfigurable Ultra-Wideband PPF

The more stages a PPF uses the larger the working band and the lower the I/Q mismatch; however, using too many stages will lead to a large voltage loss, making it a challenge for buffer design. So, for ultra-wideband application, in order to reduce voltage loss and improve I/Q quadrature performance, it is often necessary to design 2~3 PPFs parallelly, working in different frequency bands; these are selected by RF switches.

In order to achieve good orthogonality in the full pass band, the switching frequency points should also be designed at the arithmetic square root of the "high frequency points at higher frequency band" and "low frequency points at lower frequency band". The specific proof method is similar to that in Section 2.2 and is not repeated here. Just taking one PPF-I example, suppose that two three-stage PPFs working at 2 ~ ω_0 GHz and ω_0 ~ 18 GHz are used to generate ultra-wideband quadrature LO signals of 2~18 GHz. Under the premise that the minimum phase mismatch is already achieved in each PPF by applying Equation (10), the frequency of ω_0 is simulated; the results are shown in Figure 8.

It can be seen that the intersection of the two curves is at 6 GHz (that is, $\sqrt{2 \times 18}$ GHz), indicating that when the frequency band is divided into 2~6 GHz and 6~18 GHz, the overall phase mismatch is the smallest and is about 0.8°.

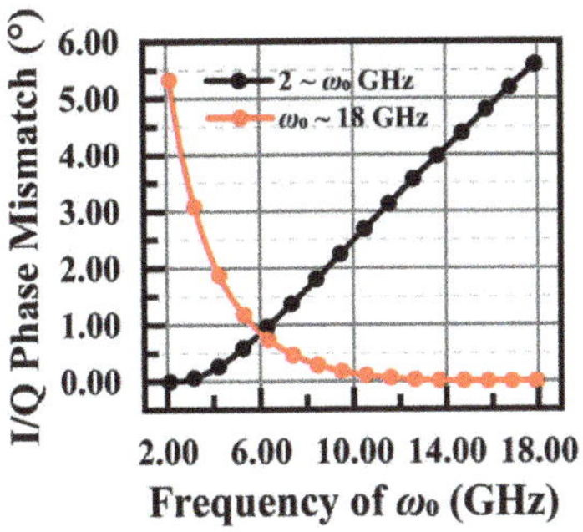

Figure 8. Phase mismatch and segmentation frequency in 2~18 G PPF.

According to the analysis above, for ultra-wideband PPF design, first segment the band with the switching frequency equaling the arithmetic square root of the "high frequency points at higher frequency band" and "low frequency points at lower frequency band"; then, decide the resonant frequencies of each band. The more bands that are divided, the better the orthogonality of the output in the whole band; however, it is worth noting that the frequency bands cannot be divided infinitely when considering the total area and symmetry of the layout design.

3. PPF Voltage Loss Model with Load Consideration

3.1. Impact of Load on PPF

The analysis presented in Section 2 is based on the traditional PPF analysis, which does not take into account the load of the PPF. In this section, the impact of the load is discussed.

To begin with, the load impact for the single-stage PPF is discussed. As shown in Figure 2a, the output of V_{I+}, without considering the load, is given in Equation (12).

$$\frac{V_{I+}}{V_{in}} = \frac{(-CR)\omega - \mathrm{i}}{(CR)\omega - \mathrm{i}} \tag{12}$$

It can be seen that the voltage gain is always equal to 1. However, as a passive network, whether in theoretical analysis or a practical test, the PPF should have losses, and the voltage gain is bound to be less than 1. For the multistage PPF itself, the input impedance of the second stage should be regarded as the load of the first stage. If it is not large enough, losses will occur. So, traditional analysis methods cannot estimate the voltage loss of the PPF.

To give a more accurate function of voltage transfer, suppose that there is a load Z_L connected to the output of PPF, as shown in Figure 9.

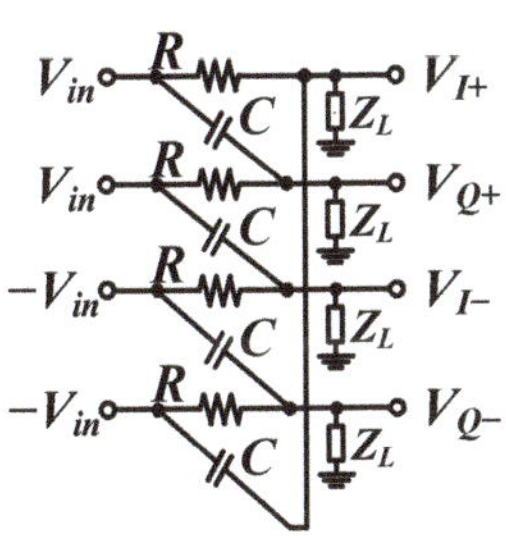

Figure 9. Single-stage PPF with load consideration.

At point X, use Kirchhoff's current law,

$$\frac{V_{I+}}{Z_L}+\frac{V_{I+}-V_{in}}{R}+\frac{V_{I+}+V_{in}}{1/j\omega C}=0 \tag{13}$$

Then, the modified transfer function is

$$\frac{V_{I+}}{V_{in}}=\frac{(-CR)\,\omega-\mathrm{i}}{(CR)\,\omega-\frac{R\mathrm{i}}{Z_L}-\mathrm{i}} \tag{14}$$

Now, the gain of the revised transfer function considering the load does not remain at one, but changes with Z_L.

The analysis above proves that the load can affect the function of output signals. Remember that in Section 2, the phase mismatch is related only to the product of the resonant frequencies of each stage, when the load is not taken into account. Next, whether this conclusion changes when the load is considered should be discussed.

Continue to take the single-stage PPF as an example. Suppose that the load is Z_L and that the output I/Q signals are

$$\frac{V_I}{V_{in}}=2\frac{(-CR)\,\omega-\mathrm{i}}{(CR)\,\omega-\frac{R\mathrm{i}}{Z_L}-\mathrm{i}} \tag{15}$$

$$\frac{V_Q}{V_{in}}=2\frac{(CR)\,\omega-\mathrm{i}}{(CR)\,\omega-\frac{R\mathrm{i}}{Z_L}-\mathrm{i}} \tag{16}$$

The ratio of the I/Q signals is

$$\frac{V_I}{V_Q}=\frac{(-CR)\,\omega-\mathrm{i}}{(CR)\,\omega-\mathrm{i}} \tag{17}$$

It can be seen that the ratio of the I/Q signals is independent of the load.

So, the conclusion is that the load will only affect the overall voltage loss of the PPF, not the orthogonality of the I/Q signals. We can continue to use the design schemes presented in Section 2 for a low I/Q mismatch. But to improve the performance of the PPF and to take it to a higher level, a deeper discussion on voltage loss should be conducted.

3.2. Input and Output Impedance Model

For the multistage PPF, the input impedance of the second stage participates in the part of the first stage as the load. So, to analyze the voltage transfer between the two stages, the input and output impedances of each stage need to be calculated first. The input impedance delivered in [10] neglected the source resistance at the input node, leading to a less accurate result due to the ignoring of the effects from the previous stage. In this section, a new analysis model for the impedance estimate is put forward to show the effect from the previous stage. The analysis model of the input impedance of the PPF is shown in Figure 10.

As shown in Figure 10, the input impedance can be calculated as a parallel circuit of a shunt starting from the resistance R with a shunt starting from capacitance C. The shunt impedance starting from R is

$$Z_R=R+\left(Z_L||\left(\frac{1}{j\omega C}+\left(Z_S||\left(R+Z_L||\left(\frac{1}{j\omega C}+\left(Z_S||\left(R+Z_L||\left(\frac{1}{j\omega C}+\left(Z_S||\left(R+Z_L||\frac{1}{j\omega C}\right)\right)\right)\right)\right)\right)\right)\right)\right)\right) \tag{18}$$

The shunt impedance starting from C is

$$Z_C=\frac{1}{j\omega C}+\left(Z_L||\left(R+\left(Z_S||\left(\frac{1}{j\omega C}+Z_L||\left(R+\left(Z_S||\left(\frac{1}{j\omega C}+Z_L||\left(R+\left(Z_S||\left(\frac{1}{j\omega C}+Z_L||R\right)\right)\right)\right)\right)\right)\right)\right)\right)\right) \tag{19}$$

The total input impedance is

$$Z_{in} = Z_C || Z_R \tag{20}$$

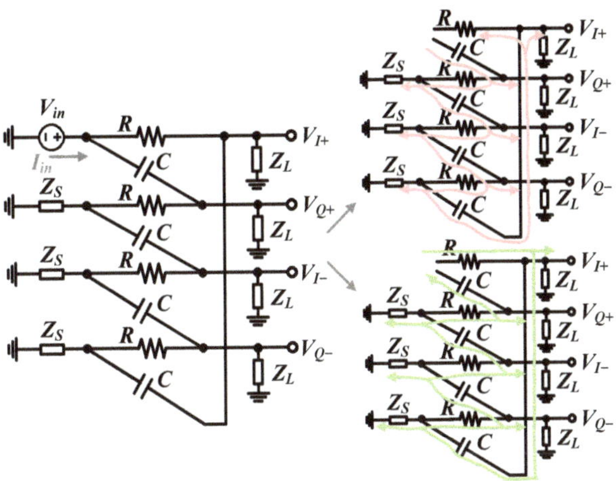

Figure 10. PPF impedance analysis model.

Figure 11 shows the comparison between the simulation results and the theoretical model results in the sweep, where Z_S = 50 Ω, Z_L = 100 Ω, R = 589 Ω, and C = 135 fF. And the curves almost overlap, indicating that the model is highly consistent with the circuit.

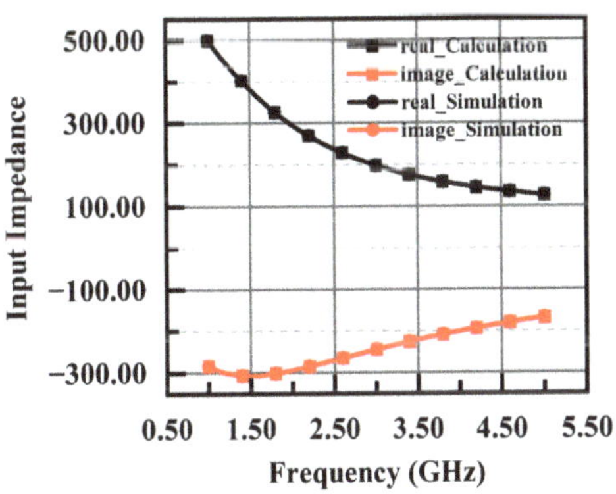

Figure 11. Input impedance of theoretical model and simulation.

For multistage PPFs like three-stage PPF-I, to calculate the input impedance of the second stage, Z_S is the impedance seen from the output node to the ground of the first stage, and Z_L is the impedance seen from the input node to the ground of the third stage, which can also be obtain by Equations (18)–(20). To simplify the analysis, when calculating Z_S and Z_L, the first and third stages can be seen as disconnected with the second, as shown in Figure 12, where Z_{S1} is the impedance of the previous stage before the PPF, and Z_{L3} is the impedance of the next stage after the PPF.

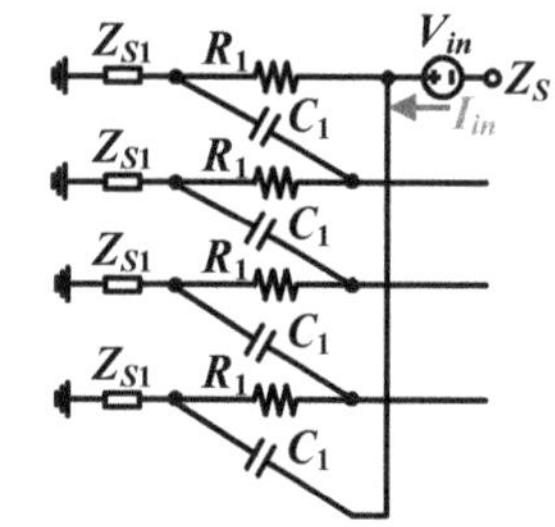

$$Z_S = \left\{ R_1 + \left(Z_{S1} \parallel \left(\frac{1}{j\omega C_1} + \left(R_1 + Z_{S1} \parallel \left(\frac{1}{j\omega C_1} + \left(R_1 + Z_{S1} \parallel \left(\frac{1}{j\omega C_1} + \left(R_1 + Z_{S1} \parallel \frac{1}{j\omega C_1} \right) \right) \right) \right) \right) \right) \right) \right\}$$

$$\parallel \left\{ \frac{1}{j\omega C_1} + \left(Z_{S1} \parallel \left(R_1 + \left(\frac{1}{j\omega C_1} + Z_{S1} \parallel \left(R_1 + \left(\frac{1}{j\omega C_1} + Z_{S1} \parallel \left(R_1 + \left(\frac{1}{j\omega C_1} + Z_{S1} \parallel R_1 \right) \right) \right) \right) \right) \right) \right) \right\}$$

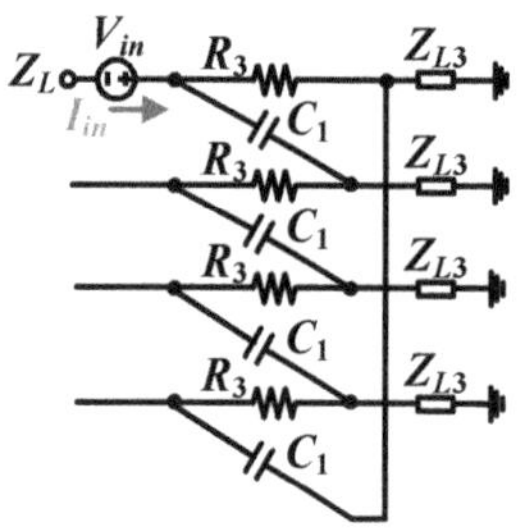

$$Z_L = \left\{ R_3 + \left(Z_{L3} \parallel \left(\frac{1}{j\omega C_3} + \left(R_3 + Z_{L3} \parallel \left(\frac{1}{j\omega C_3} + \left(R_3 + Z_{L3} \parallel \left(\frac{1}{j\omega C_3} + \left(R_3 + Z_{L3} \parallel \frac{1}{j\omega C_3} \right) \right) \right) \right) \right) \right) \right) \right\}$$

$$\parallel \left\{ \frac{1}{j\omega C_3} + \left(Z_{L3} \parallel \left(R_3 + \left(\frac{1}{j\omega C_3} + Z_{L3} \parallel \left(R_3 + \left(\frac{1}{j\omega C_3} + Z_{L3} \parallel \left(R_3 + \left(\frac{1}{j\omega C_3} + Z_{L3} \parallel R_3 \right) \right) \right) \right) \right) \right) \right) \right\}$$

Figure 12. Calculation of ZS and ZL for second stage.

The result will be slightly different from that of the circuit simulation, as the multistage PPF has much more complicated series or parallel branches; it can also play a part as a great estimation tool in the whole design.

Note that although the model is given by input impedance analysis, it can also be used to calculate the output impedance. The process is similar; so, it is not repeated here.

3.3. Voltage Loss Estimation of Multistage PPF

Reference [10] gives the equations of voltage loss for a certain stage of the PPF. In this section, a more universal solution is proposed for calculating the loss between any two stages, and the analysis model is shown in Figure 13.

In the PPF shown in Figure 13, the second stage can be regarded as the load of the first stage, and the voltage transfer function is shown in Equations (21) and (22).

$$\frac{V_{I1+}}{V_{in}} = \frac{(-C_1 R_1 Z_1)\,\omega - Z_1}{(-C_1 R_1 Z_1\,\mathrm{i})\,\omega - R_1 - Z_1} \tag{21}$$

$$\frac{V_{I2+}}{V_{in}} = \frac{(-C_1 C_2 R_1 R_2 Z_1 Z_2)\,\omega^2 + (-C_1 R_1 Z_1 Z_2 - C_2 R_2 Z_1 Z_2)\,\omega - Z_1 Z_2}{(C_1 C_2 R_1 R_2 Z_1 Z_2)\,\omega^2 + (-C_1 R_1 R_2 Z_1\,\mathrm{i} - C_2 R_1 R_2 Z_2\,\mathrm{i} - C_1 R_1 Z_1 Z_2\,\mathrm{i} - C_2 R_2 Z_1 Z_2\,\mathrm{i})\,\omega - R_1 R_2 - R_1 Z_2 - R_2 Z_1 - Z_1 Z_2} \tag{22}$$

Z_1 is just the input impedance of the second stage, which was discussed in Section 3.2. As for Z_2, for the two-stage PPF Z_2 is just the load impedance connected to the output end; in the case of three or more stages, the analysis of Z_2 should also refer to the process of Z_1.

In order to verify the reliability of the model with examples, the theoretical results from the analysis above and circuit simulation of a two-stage PPF are shown in Figure 14. It can be seen that the calculated results of the theoretical model are slightly larger than the simulation results of the circuit, indicating that the actual voltage transfer model of the PPF should be more complex. However, this is a feasible model for roughly estimating the output voltage; it can guide designers in setting the indicators of the whole system and can facilitate the estimation of the circuit performance at the beginning of circuit design.

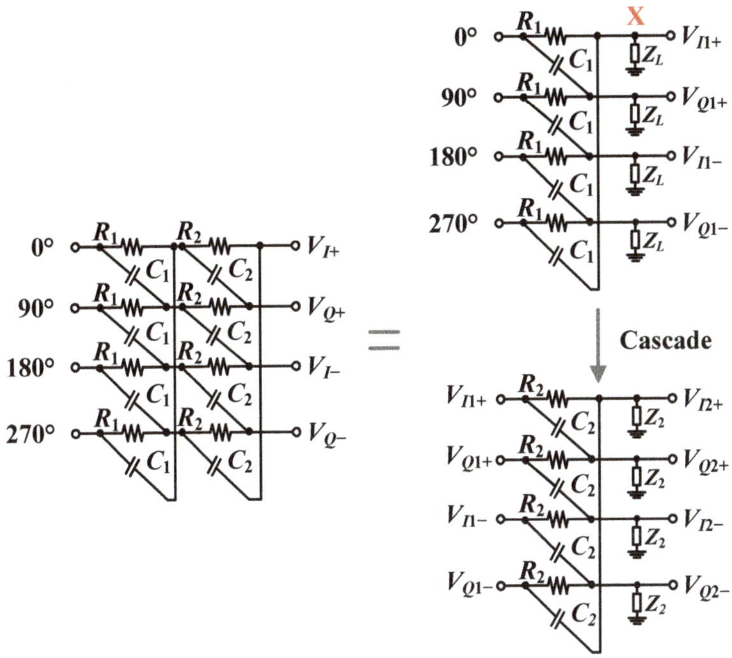

Figure 13. PPF Interstage loss estimation Model.

Another discovery is that, according to Equations (17)–(22), the individual values of the resistance and capacitance of each stage play a significant role in the voltage loss. Therefore, although the individual values of resistance and capacitance have no influence on the I/Q mismatch, they should still be well designed to minimize the total voltage loss, according to the premise that their production remains unchanged.

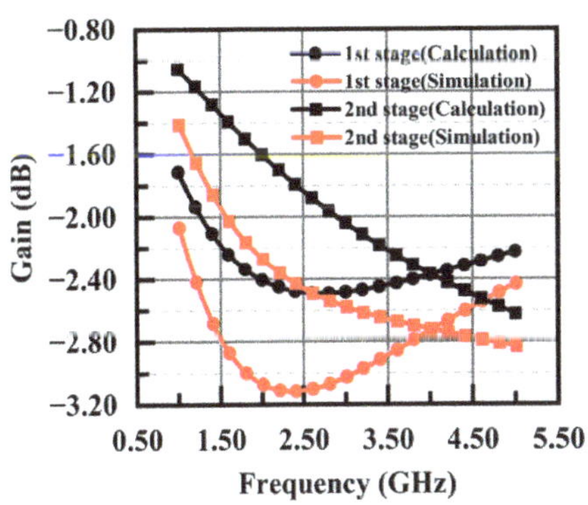

Figure 14. Voltage loss of PPF calculated by theoretical model and circuit simulation.

4. Circuit Design: 2~8 GHz Reconfiguration PPF

According to the analysis of the phase/amplitude mismatch and the voltage loss in Section 3, a universal PPF design model can be generated. In this section, a design example of a 2–8 GHz PPF is used to detail this model.

4.1. PPF Structure Determination

As shown in Figure 2, there are two PPF structures: PPF-I, which theoretically has no magnitude mismatch, and PPF-II, which theoretically has no phase mismatch. However, if the PPF is required to have a phase mismatch that is as low as possible, it should be designed like PPF-I, as the deviation of the perfectly orthogonal frequency points caused by parasitic effects can be corrected by adjusting the designed resonant frequencies in the circuit. Similarly, if the PPF is required to have a magnitude noise that is as low as possible, it should be designed like PPF-II.

In this design, because a low phase mismatch is desired, a structure with two inputs connected together is applied.

4.2. Working Band Segmentation (Reconfiguration Modularization)

As discussed in Section 2.3, to achieve a low I/Q mismatch for the ultra-wideband application, more than one PPF working in different frequency bands should be used, and the switching frequency should be the arithmetic square root of the "high frequency points at higher frequency band" and "slow frequency points at lower frequency band".

Therefore, for the frequency range of 2~8 GHz, two PPFs working at 2~4 GHz and 4~8 GHz, respectively, are used to form the overall circuit. Theoretically, as shown in Figure 15, the maximum phase mismatch of the whole band is about 0.22°. Band switching can be achieved by an RF switch like a transmission gate.

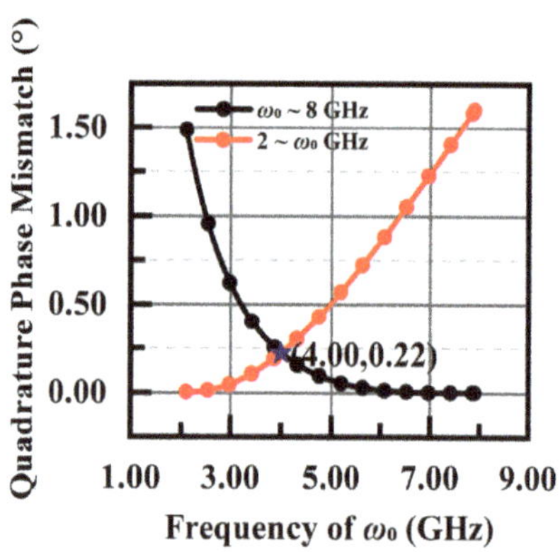

Figure 15. Phase mismatch and segmentation frequency in 2~8 GHz PPF.

4.3. Resonant Frequency Determination of Each Stage of Each PPF

In this design, a three-stage PPF is used in each band for voltage loss and phase mismatch trade-off. According to the discussion in Section 2.2, when $\omega_2 = \sqrt{\omega_1\omega_3}$, the I/Q quadrature performance is the best. So, the resonant frequencies for a 2~4 G PPF are 2 GHz, 2.83 GHz, and 4 GHz; the resonant frequencies for a 4~8 GHz PPF are 4 GHz, 5.66 GHz, and 8 GHz.

4.4. R&C Determination Based on Voltage Loss

After obtaining the three resonant frequencies, the product of the resistance and capacitance at each stage can be obtained according to $\omega = 1/RC$. According to the analysis in Section 3.3, the respective values of the resistors and capacitors play an important role in the amplitude of the output voltage of the PPF. To maximize the output voltage of the PPF, it is necessary to determine the specific values of resistance and capacitance at each stage, according to the input impedance of the output buffer and the output impedance of the input buffer.

4.5. Layout Design

In the traditional design, the PPF layout is often drawn as a circuit schematic diagram, as shown in Figure 4, which leads to the long route between the last capacitor and the first resistance. At high frequency, long wires may have a significant parasitic inductance which is almost near the order of pH/μm. So, in the traditional layout, the length of the long wire is from tens of μm to hundreds of μm, depending on the width and length of the selected resistance and capacitor. Under the influence of this parasitic inductance and other parasitic capacitance, the required resonant frequency will be shifted.

In order to reduce this effect, the improvement proposed in this paper is to rearrange the three-stage resistors and capacitors, as shown in Figure 16; thus, the four output signals have the same routing length, as far as possible, and the parasitic influence of each route is the same, as far as possible; the overall layout is more symmetrical, reducing the influence of parasitism on the I/Q quadrature performance. And for a multistage PPF application, this routing scheme is also very convenient for cascading.

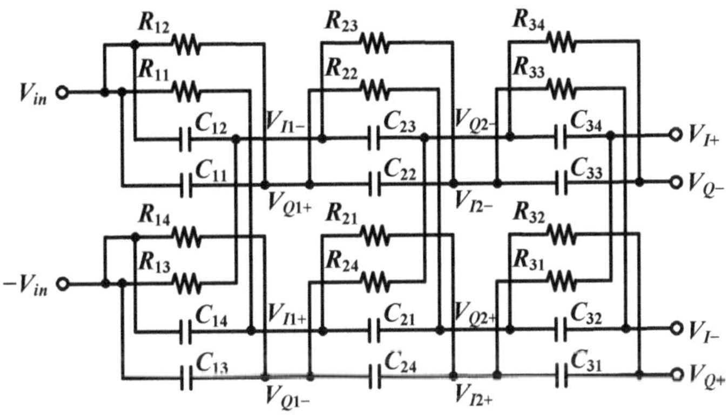

Figure 16. Improved routing scheme of three-stage PPF layout.

As the parasitic resistance and capacitance will be connected in parallel or series with the intrinsic resistance and capacitance, resulting in the shifting of the resonant frequency, it is necessary to check whether the resonant frequencies of the three stages are too far from the desired one. If so, it is necessary to return to the schematic and adjust the resistance and capacitance.

5. Results and Discussions

The layout of the final 2~8 GHz three-stage PPF in a 55 nm process is shown in Figure 17. The upper and lower parts work at 2~4 GHz and 4~8 GHz, respectively, and are switched by transmission gates. The area is 174 × 145 μm^2. The resistors and capacitors used are shown in Table 1.

Table 1. Parameters used in PPF.

2~4 GHz		4~8 GHz	
R_1/Ω	110.911	R_1/Ω	188.903
C_1/fF	727.306	C_1/fF	185.883
R_2/Ω	150.307	R_2/Ω	162.561
C_2/fF	342.712	C_2/fF	180.931
R_3/Ω	198.416	R_3/Ω	133.004
C_3/fF	199.426	C_3/fF	128.639

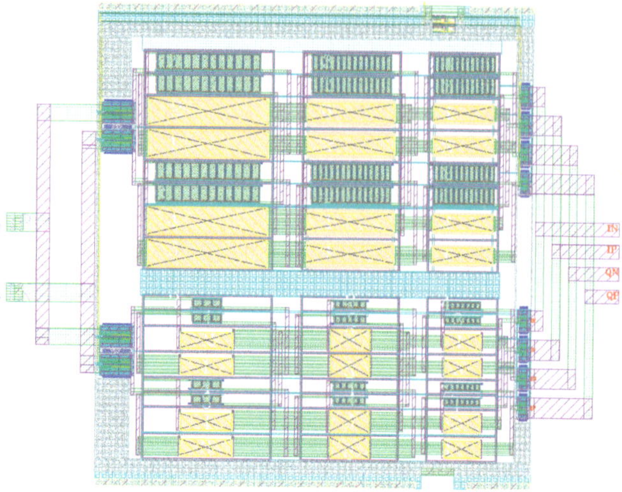

Figure 17. Layout of PPF.

Figures 18 and 19 show simulation results of post-simulation. With the output connected to an RF amplifier, whose input impedance is 30.319–292.34j throughout the 2~8 GHz frequency band, the quadrature phase mismatch is ≤0.2439°, the amplitude mismatch is ≤0.098 dB, and the voltage loss is ≤17.7 dB, in post-simulation.

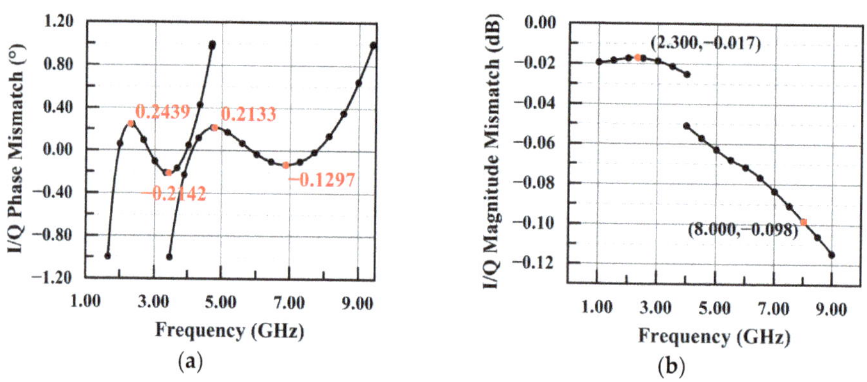

Figure 18. I/Q mismatch for (**a**) phase mismatch and (**b**) magnitude mismatch.

The parasitic parameter extraction was performed on the circuit layout, followed by Monte Carlo simulations and statistical fitting, as illustrated in Figure 20.

According to Figure 20, in the case of I/Q phase mismatch, the majority of samples are within the range of 0.225° to 0.325°, with a mean of 0.275°. Post-fitting analysis reveals an approximate adherence to a Gaussian distribution, with a standard deviation (σ) of approximately 0.0386°, and the 3σ interval of the Gaussian distribution is approximately 0.13° to 0.43°. Regarding the I/Q amplitude mismatch, most samples fall within the range of 0.095 dB to 0.105 dB, with a mean of 0.100 dB. After fitting, an approximate adherence to a Gaussian distribution is observed, with a σ of approximately 0.00512 dB, and the 3σ interval of the Gaussian distribution is approximately 0.079 dB to 0.121 dB. The foregoing analysis suggests a robustness in both the phase and the amplitude against variations in PVT and manufacturing.

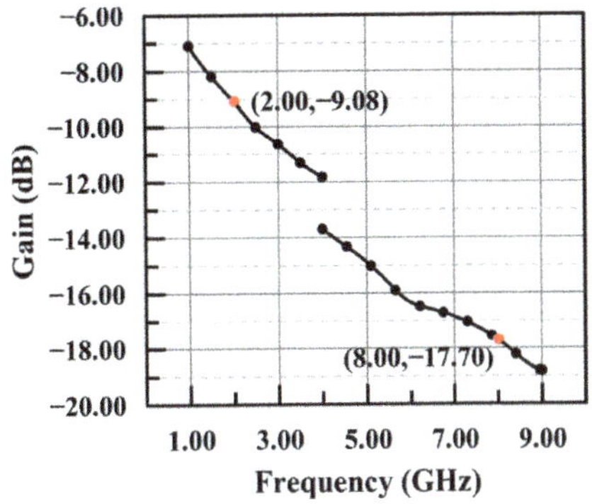

Figure 19. Voltage gain (loss).

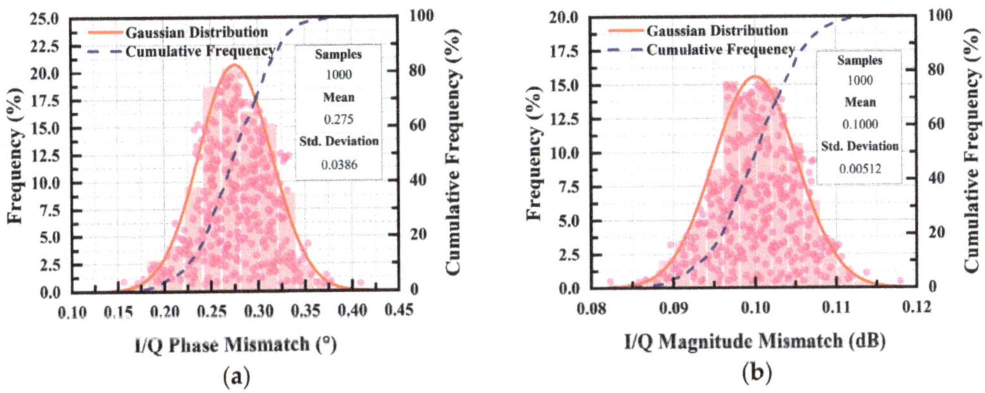

Figure 20. I/Q mismatch in Monte Carlo sampling for (**a**) phase mismatch and (**b**) magnitude mismatch.

The comparisons with other works in the past filter years for PPFs without buffers are shown in Table 2, indicating that the PPF designed by the method presented in this paper can achieve better performance, which has a very small I/Q mismatch with a moderated voltage loss.

Table 2. Comparison table.

	This Work	[11]	[12]	[13]
Technology	55 nm CMOS	0.15 μm AlGaAs/InGaAs	65 nm CMOS	65 nm CMOS
Frequency (GHz)	2~8	32.5~34/14.7~15.2	26.5~29.5	33~39
I/Q Phase Mismatch (°)	$\leq$0.2439	$\leq$1	$\leq$1.77	$\leq$1.1
I/Q Magnitude Mismatch (dB)	$\leq$0.098	$\leq$0.5	$\leq$0.27	$\leq$0.28
Voltage Loss (dB)	≤ -17.70	--	≤ -13	≤ -17.3
Area (μm^2)	174 × 145	--	100	140

Based on this PPF, the four-phase generation network can achieve a relatively low I/Q mismatch and power consumption and can improve the overall performance of the circuit with a well-designed single-to-differential circuit and RF amplifier buffer.

6. Conclusions

This paper discusses a design and estimation scheme for a wideband reconfigurable low I/Q mismatch passive polyphase filter (PPF) in a 2~8 GHz frequency band. The relationship between the designed parameters and the PPF performance is proposed, which indicates that the key to the low mismatch lies in the proper switching frequency of the reconfigurable band and the proper resonant frequency of each stage, and the key to the low loss lies in the moderate values of resistance and capacitance, respectively. Based on this model, a 2~8 GHz band PPF is designed in a 55 nm CMOS process. The post-simulation results show that the designed PPF achieves a phase mismatch of within 0.2439° and an amplitude mismatch of within 0.098 dB within the whole frequency band. Through Monte Carlo sampling and statistical distribution fitting, the circuit demonstrates strong robustness against PVT variations and manufacturing changes. Under the load of 30.319–292.34j, it can achieve a loss of within 17.7 dB, which reduces the design difficulty of the RF amplifier buffer. The area of the whole PPF module is 0.174×0.145 mm^2.

Author Contributions: Conceptualization, Y.W. (Yingze Wang) and X.L.; data curation, Y.W. (Yuanze Wang) and F.H.; formal analysis, Y.W. (Yingze Wang), X.L. and Y.W. (Yuanze Wang); funding acquisition, X.W. and X.L.; investigation, F.H., Z.L., X.W. and X.L.; methodology, Y.W. (Yingze Wang) and Q.Q.; project administration, X.W., X.L. and Z.L.; supervision, X.W., X.L. and Z.L.; validation, Y.W. (Yingze Wang) and F.H.; visualization, X.L. and Z.L.; writing—original draft, Y.W. (Yingze Wang) and Y.W. (Yuanze Wang); writing—review and editing, Y.W. (Yingze Wang) and X.L. All authors have read and agreed to the published version of the manuscript.

Funding: This work was supported in part by Beijing Institute of Technology Research Fund Program for Young Scholars under Grant XSQD-202105006, in part by the National Natural Science Foundation of China under Grant 62101038.

Data Availability Statement: The original contributions presented in the study are included in the article, further inquiries can be directed to the corresponding authors.

Conflicts of Interest: The authors declare no conflicts of interest.

References

1. Haddad, F.; Rahajandraibe, W.; Zaid, L.; Frioui, O.; Bouchakour, R.; IEEE. Design of Radio Frequency Passive Polyphase Filter for 2.4 GHz Wireless Communication Applications. In Proceedings of the 2009 IEEE 10th Annual Wireless and Microwave Technology Conference, Clearwater, FL, USA, 20–21 April 2009; pp. 142–145. [CrossRef]
2. Kulkarni, S.; Zhao, D.; Reynaert, P. Design of an optimal layout polyphase filter for millimeter-wave quadrature lo generation. *IEEE Trans. Circuits Syst. II Express Briefs* **2013**, *60*, 202–206. [CrossRef]
3. Sato, L.; Shouno, K.; Tanimoto, H.; Muto, C.; Yoshino, M.; Moriyama, S.; Takahashi, C. Measurement of the Frequency Characteristics of an RC Polyphase Filter Using 4-phase Square Waves. In Proceedings of the 2022 37th International Technical Conference on Circuits/Systems, Computers and Communications (ITC-CSCC), Phuket, Thailand, 5–8 July 2022; pp. 1–4. [CrossRef]
4. Sahoo, A.K.; Rawat, K. Wideband Analog Polyphase Network for Harmonic Cancellation of UHF Power Amplifier. In *IEEE Transactions on Microwave Theory and Techniques*; IEEE: New York, NY, USA, 2022. [CrossRef]
5. Bautista, M.; Dutkiewicz, E.; Huang, X.; Nguyen, D.; Zhu, F. Quadrature Broadband Phase Shift Generation Using Passive RC Polyphase Filter for RF Front-End. In Proceedings of the 2016 16th International Symposium on Communications and Information Technologies (ISCIT), Qingdao, China, 26–28 September 2016; pp. 597–601. [CrossRef]
6. Sethi, A.; Akbar, R.; Hietanen, M.; Rahkonen, T.; Pärssinen, A. A 26GHz to 34GHz Active Phase Shifter with Tunable Polyphase Filter for 5G Wireless Systems. In Proceedings of the 2021 16th European Microwave Integrated Circuits Conference (EuMIC), London, UK, 3–4 April 2022; pp. 301–304. [CrossRef]
7. Xiang, X.; Gai, W.; Shi, L.; He, A.; Sheng, K. An 8–12GHz 0.92° Phase Error Quadrature Clock Generator Based on Two-Stage Poly Phase Filter with Intermediate Point Compensation. In Proceedings of the 2019 IEEE International Symposium on Circuits and Systems (ISCAS), Sapporo, Japan, 26–29 May 2019; pp. 1–5. [CrossRef]
8. Momeni, M.; Moezzi, M. A Low Loss and Area Efficient RC Passive Poly Phase Filter for Monolithic GHz Vector-Sum Circuits. *IEEE Trans. Circuits Syst. II Express Briefs* **2019**, *66*, 1134–1138. [CrossRef]
9. Kim, C.-W.; Lee, S.-G. A 5.25-GHz image rejection RF front-end receiver with polyphase filters. *IEEE Microw. Wireless Compon. Lett.* **2006**, *16*, 302–304.
10. Kaukovuori, J.; Stadius, K.; Ryynanen, J.; Halonen, K.A.I. Analysis and Design of Passive Polyphase Filters. *IEEE Trans. Circuits Syst. I Regul. Pap.* **2008**, *55*, 3023–3037. [CrossRef]

11. Su, J.-Y.; Tseng, S.-C.; Meng, C.; Wu, P.-Y.; Lee, Y.-T.; Huang, G.-W. *Ka/Ku*-Band pHEMT Gilbert Mixers with Polyphase and Coupled-Line Quadrature Generators. *IEEE Trans. Microw. Theory Tech.* **2009**, *57*, 1063–1073. [CrossRef]
12. Tang, M.; Zhang, Y.; Pang, J.; Shirane, A.; Okada, K. A 28GHz Area-Efficient CMOS Vector-Summing Phase Shifter Utilizing Phase-Inverting Type-I Poly-Phase Filter for 5G New Radio. In Proceedings of the ESSCIRC 2022-IEEE 48th European Solid State Circuits Conference (ESSCIRC), Milan, Italy, 19–22 September 2022; pp. 333–336. [CrossRef]
13. Li, Y.; Duan, Z.; Lv, W.; Pan, D.; Xie, Z.; Dai, Y.; Sun, L. A 32–40 GHz 7-bit CMOS Phase Shifter with 0.38 dB/1.6° RMS Magnitude/Phase Errors for Phased Array Systems. In *2020 IEEE Radio Frequency Integrated Circuits Symposium (RFIC)*; IEEE: New York, NY, USA, 2020.

electronics

MDPI

Communication

Design Techniques for L-C-L T-Type Wideband CMOS Phase Shifter with Suppressed Phase Error

Seongjin Jang [1] and Changkun Park [1,2,*]

[1] Department of Electric Engineering, Soongsil University, 369, Sangdo-ro, Dongjak-gu, Seoul 06978, Republic of Korea; jsj0613@soongsil.ac.kr

[2] Department of Intelligent Semiconductors, Soongsil University, 369, Sangdo-ro, Dongjak-gu, Seoul 06978, Republic of Korea

* Correspondence: pck77@ssu.ac.kr; Tel.: +82-828-7166

Abstract: In this study, we designed a K-band CMOS switch-type phase shifter. Equivalent circuits of shift and pass modes were analyzed to minimize phase errors in a wide frequency range. In particular, the impedance inside the equivalent circuit of the pass mode was analyzed to derive a frequency region in which the equivalent circuit of the pass mode becomes an L-C-L structure. Based on the fact that equivalent circuits in shift and pass modes can be regarded as L-C-L structures beyond a specific frequency, a design methodology of the wideband phase shifter was proposed through slope adjustment of the phase according to the frequency of each of the two modes. To verify the feasibility of the proposed design methodology, a 20°-bit phase shifter was designed through a 65 nm RFCMOS process. As a result of the measurement at 21.5 GHz to 40.0 GHz, the phase error was within 0.87°.

Keywords: CMOS; phase error; phase shifter; wideband

1. Introduction

With the introduction of 5G mobile communication, research related to beamforming systems has recently been actively conducted [1–4]. Accordingly, millimeter-wave (mm-Wave) circuits constituting the beamforming system are also developing. Recently, research has been active in securing more frequency bands with one beamforming system by securing wideband characteristics as well as improving the performance of mm-Wave circuits that make up the beamforming system [5–10].

With the development of such a beamforming system, research on a phase shifter for antenna beam control is also attracting attention [11–13]. Such a phase shifter is divided into an active type using a vector sum and a passive type using a filter consisting of a passive device and switch. Active phase shifters have the advantage of fine-tuning the phase and amplifying the signal, while only one-way signal paths are possible [14–21]. On the other hand, in the case of a passive phase shifter, there is a disadvantage that insertion loss occurs due to power loss caused by passive devices, but there is an advantage that a bidirectional signal path is possible [22–25]. The passive phase shifter consists of several bits that can control the phase, and each bit has a different amount of phase control function. The advantages and disadvantages of each phase shift structure have been summarized in previous studies [26]. These phase shifters, like other mm-Wave circuits [5,9], require wideband characteristics to be secured for the application of beamforming systems that can support multiple frequency bands [11,25].

In this study, a method of designing a unit bit of a phase shifter that can secure wideband characteristics in a passive phase shifter consisting of several bits and suppress phase error at the same time was proposed. The proposed design methodology relates to the L-C-L T-type structure, and a method of maintaining phase differences in a wide range of frequency bands through impedance analysis in pass and shift modes was proposed. The proposed design methodology was applied to the phase shifter fabricated with the 65 nm RFCMOS process, and its feasibility was confirmed through the measurement results.

2. Proposed Design Method of the Wideband L-C-L T-Type Unit Bit for Phase Shifters

Here, in order to explain the proposed design methodology of a wideband phase shifter with a suppressed phase error, the equivalent circuits in shift and pass mode were first analyzed. After considering the simplified equivalent circuit for each mode, a design technique for a wideband phase shifter with suppressed phase errors was presented using the derived equivalent circuits.

2.1. Analysis of Equivalent Circuits by Mode in T-Type Structure

Figure 1 shows the schematic and its equivalent circuits of the unit bit of a wideband phase shifter for the application of the proposed design methodology. The unit bit of the phase shifter operates in shift and pass modes, respectively, depending on the on- and off-states of the transistor acting as a switch. Figure 1b,c show equivalent circuits in shift and pass modes of the unit bit, respectively.

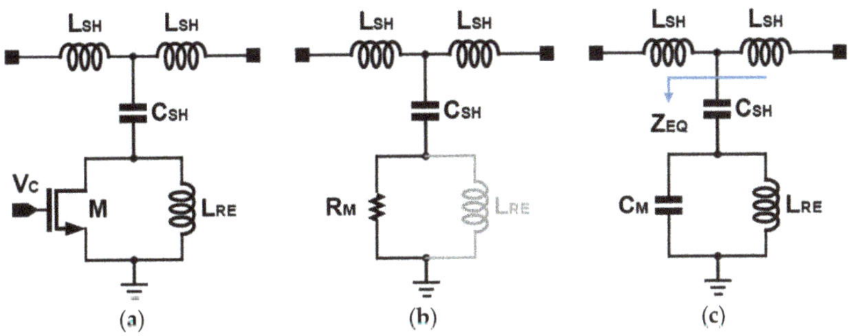

Figure 1. L-C-L T-type structure of unit bit for phase shifter: (**a**) schematic of the used T-type structure, and equivalent circuits for the (**b**) shift and (**c**) pass modes.

In the shift mode shown in Figure 1b, the transistor M is turned on, and for convenience of analysis, it is assumed equivalent to the on-resistance, R_M of the transistor. Assuming that the impedance of R_M is small enough, the impedance by L_{RE} connected in parallel with R_M can be ignored. In this case, the equivalent circuit of the shift mode is in the form of a low-pass filter composed of a T-type L_{SH}-C_{SH}-L_{SH}.

On the other hand, in the case of the pass mode, M is turned off, and for the convenience of analysis, it is assumed equivalent to C_M due to parasitic capacitances of the transistor M, as shown in Figure 1c. In this case, unlike the shift mode, direct equalization with the L-C-L structure is difficult. Therefore, the Z_{EQ} of Figure 1c was developed as follows.

$$Z_{EQ} = \frac{1 - \omega^2 L_{RE}(C_{SH} + C_M)}{j\omega C_{SH}(1 - \omega^2 L_{RE} C_M)} \tag{1}$$

Here, for convenience of analysis, it was assumed that the parasitic resistance was small enough to be negligible. The frequencies at which Z_{EQ} becomes 0 and infinite using Equation (1) are calculated as follows.

$$\omega_{Z_{EQ}=0} = \frac{1}{\sqrt{L_{RE}(C_{SH}+C_M)}}, \quad \omega_{Z_{EQ}=\infty} = \frac{1}{\sqrt{L_{RE}C_M}} \tag{2}$$

Here, $\omega_{Z_{EQ}=0}$ and $\omega_{Z_{EQ}=\infty}$ denote frequencies at which Z_{EQ} becomes 0 and infinity, respectively. Figure 2 shows the simulation results of the resistance and capacitance of Z_{EQ} according to frequency. Since the simulation was performed using models of actual devices, there is a slight difference from the results of the analyzed equations, but the tendency is the same as that of the equations. The main difference between equations and simulation is that parasitic resistances were ignored in equations, but were considered in simulation.

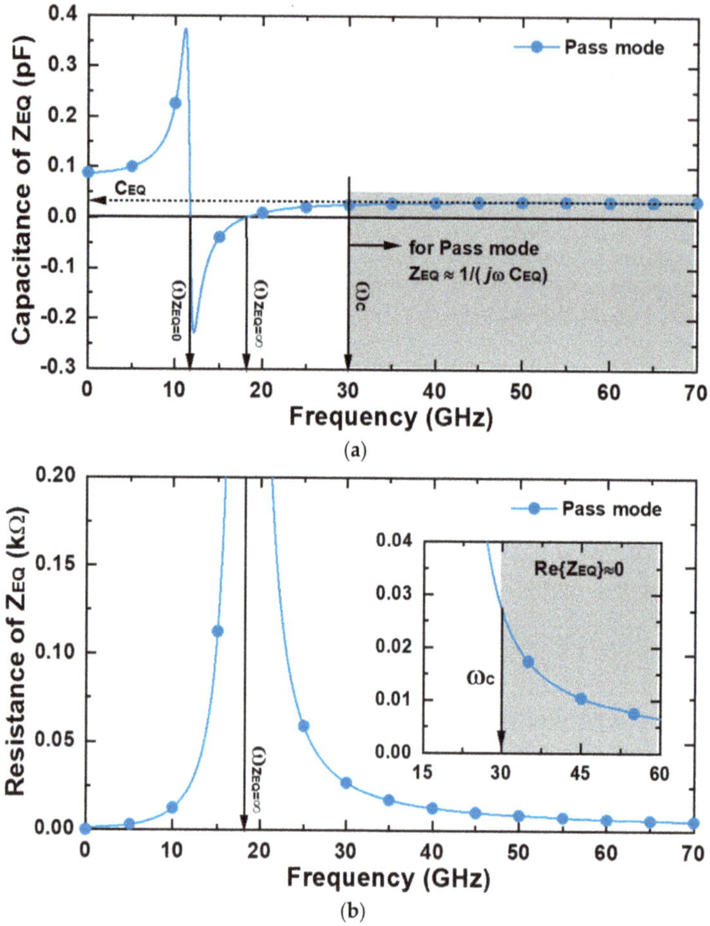

Figure 2. Simulated conceptual Z_{EQ} according to operating frequency: (**a**) capacitance and (**b**) resistance.

The capacitance of Z_{EQ} shown in Figure 2a has a value of 0 at $\omega_{Z_{EQ}=0}$ and $\omega_{Z_{EQ}=\infty}$. As the operating frequency gradually increases from $\omega_{Z_{EQ}=0}$, the capacitance of Z_{EQ} converges to a specific value, and the converging capacitance is expressed as C_{EQ}. At this time, the frequency at which the capacitance of Z_{EQ} begins to be regarded as C_{EQ} was defined as ω_C. On the other hand, the resistance shown in Figure 2b has the highest value at $\omega_{Z_{EQ}=\infty}$, which is determined by the L_{RE} and C_M as shown in Equation (2). As the operating frequency increases from $\omega_{Z_{EQ}=\infty}$, the resistance gradually decreases, and eventually approaches 0 Ω. In addition, even near ω_C set from Figure 2a, the resistance of Z_{EQ} can be considered 0 Ω. As a result, in the frequency region higher than ω_C, the resistance and capacitance of Z_{EQ} are 0 Ω and C_{EQ}, respectively, so Z_{EQ} can be expressed as follows.

$$Z_{EQ} \cong \frac{1}{j\omega C_{EQ}} \quad for \ \ \omega > \omega_C \tag{3}$$

Therefore, in the case of pass-mode, a circuit consisting of C_{SH}, C_M, and L_{RE} may be represented by C_{EQ} in the frequency region higher than ω_C. This allows an equivalent circuit in pass mode to be T-type L_{SH}-C_{EQ}-L_{SH}, similar to shift mode, in frequency regions higher than ω_C. Figure 3 shows equivalent circuits of shift and pass modes in frequency regions higher than ω_C.

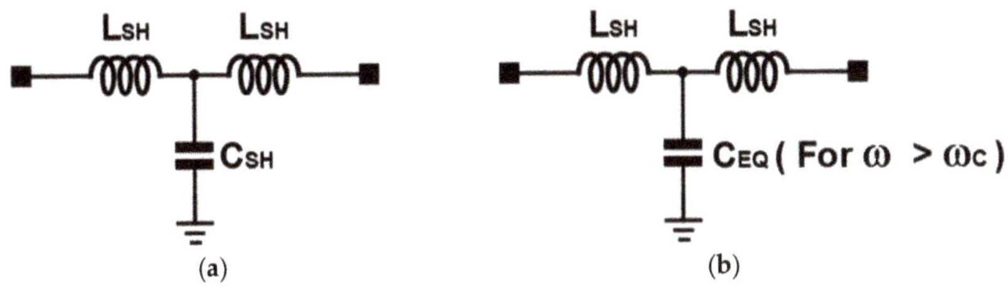

Figure 3. Equivalent circuits in frequency regions above ω_C: (**a**) shift and (**b**) pass modes.

2.2. Proposed Design Methodology for Wideband Phase Shifter with Suppressed Phase Error

Figure 4 is a conceptual diagram for explaining the proposed design methodology of a wideband phase shifter with suppressed phase errors using the previously derived equivalent circuit for each mode. In the case of an equivalent circuit in shift mode, it has a structure of L_{SH}-C_{SH}-L_{SH} regardless of the operating frequency, so the phase in shift mode in Figure 4 shows linear characteristics according to the frequency. On the other hand, in the case of an equivalent circuit in pass mode, it can only be regarded as the structure of L_{SH}-C_{EQ}-L_{SH} in the higher operating frequency region than ω_C, so the phase in pass mode in Figure 4 shows linear characteristics according to the frequency after ω_C.

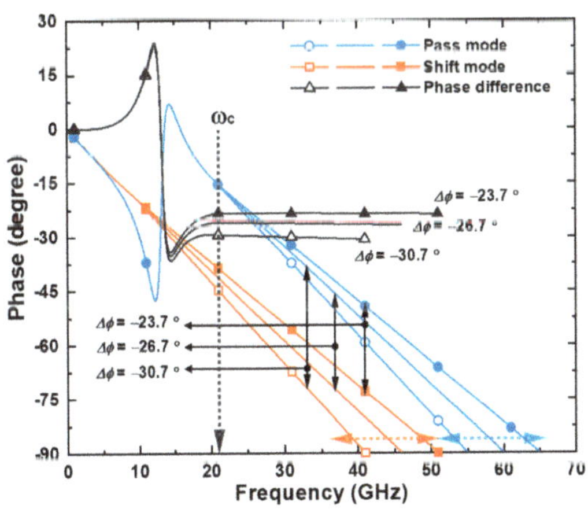

Figure 4. Conceptual diagram for securing wideband characteristics with suppressed phase error.

That is, in the frequency regions after ω_C, for both shift and pass modes, the phase according to the frequency maintains a linear characteristic. Therefore, in the frequency region after ω_C, if the slopes of the phases according to the frequency in the shift and pass modes are set to be the same, the phase difference between shift and pass modes can be maintained to be the same in the frequency region after ω_C. When the slopes of the shift and pass modes are set the same as each other, the phase difference between the two modes also increases as the absolute value of the slope in the two modes increases.

As a result of qualitative analysis, wideband characteristics can be secured by equalizing the phase slopes according to the frequency of the two modes. In addition, the desired phase difference can be secured by adjusting the equalized slope.

In order to quantify such a qualitative analysis, the phase of each mode in the frequency region after ω_C can be calculated as follows.

$$\varphi_{Shift} = \tan^{-1}\left(-\frac{2\omega L_{SH} - \omega^3 L_{SH}^2 C_{SH}}{Z_0(1-\omega^2 L_{SH} C_{SH})}\right) \tag{4}$$

$$\varphi_{Pass} = \tan^{-1}\left(-\frac{2\omega L_{SH} - \omega^3 L_{SH}^2 C_{EQ}}{Z_0(1-\omega^2 L_{SH} C_{EQ})}\right) \tag{5}$$

where φ_{Shift} and φ_{Pass} are phases of the shift and pass modes, respectively. In addition, we assumed that the termination impedance Z_0 is 50 Ω. Here, for convenience of analysis through equations, it was assumed that parasitic resistances including R_M were small enough to be negligible. In this study, instead of directly setting the phase slope according to the frequency, the desired slope for each mode was secured by adjusting the frequency at which the phase in each mode becomes −90°. In order to set the slope of the phase in this manner, the frequency at which the phase becomes −90° for each mode is considered as follows.

$$\begin{aligned} \omega_{Shift-90} &= \frac{1}{\sqrt{L_{SH}C_{SH}}} \\ \omega_{Pass-90} &= \frac{1}{\sqrt{L_{SH}C_{EQ}}} \end{aligned} \tag{6}$$

where $\omega_{Shift-90}$ and $\omega_{Pass-90}$ are frequencies when the phase becomes −90° in shift and pass modes, respectively.

As a result of quantitative analysis of the proposed design methodology, if the frequency with a phase of −90° in each mode is adjusted through Equation (6), a wideband phase shifter with a suppressed phase error can be secured.

3. Design Examples of the Phase Shifter Using Proposed Design Methodology

In this study, in order to verify the effectiveness of the proposed design methodology, several unit bits of the phase shifter were first designed through simulation.

Here, if Equation (6) and Figure 4 are analyzed again, in the case of the unit bit of the L-C-L structure of a wideband phase shifter with a large phase difference, the absolute value of the phase slope according to frequency must increase. This means that $\omega_{Shift-90}$ and $\omega_{Pass-90}$ of Equation (6) should decrease, and $L_{SH}C_{SH}$ and $L_{SH}C_{EQ}$ should increase. As such, when the required inductance and capacitance increase for a large phase difference, the power loss and the size of the integrated circuit also increase accordingly. Therefore, the wideband unit bit with suppressed phase error using the T-type L-C-L structure proposed in this study is relatively more suitable for securing a small phase difference. For this reason, in this study, as shown in Figure 5, unit bits for phase shifters of 5°, 10°, 15°, and 20° were designed to verify the proposed structure through simulation.

In this study, the 65 nm RFCMOS process which provides eight metal layers was used to design the phase shifter. The schematics for the simulation results of Figure 5 are all the same as Figure 1a. The device values used for each bit to obtain the phase shift of 5°, 10°, 15°, and 20° shown in Figure 5 are summarized in Table 1. When designing for each bit to obtain simulation results, the effects of metal lines, inductors, and test pads were all considered through electromagnetic (EM) simulation to secure the accuracy of the simulation results. Spiral inductors are designed using the top metal layer to minimize loss due to the Silicon substrate. In addition, the gate voltages of the transistor were 1 V and 0 V, respectively, for the on and off states of the transistor. The simulation results in Figure 5 were conducted at a temperature of 25 °C in a typical corner.

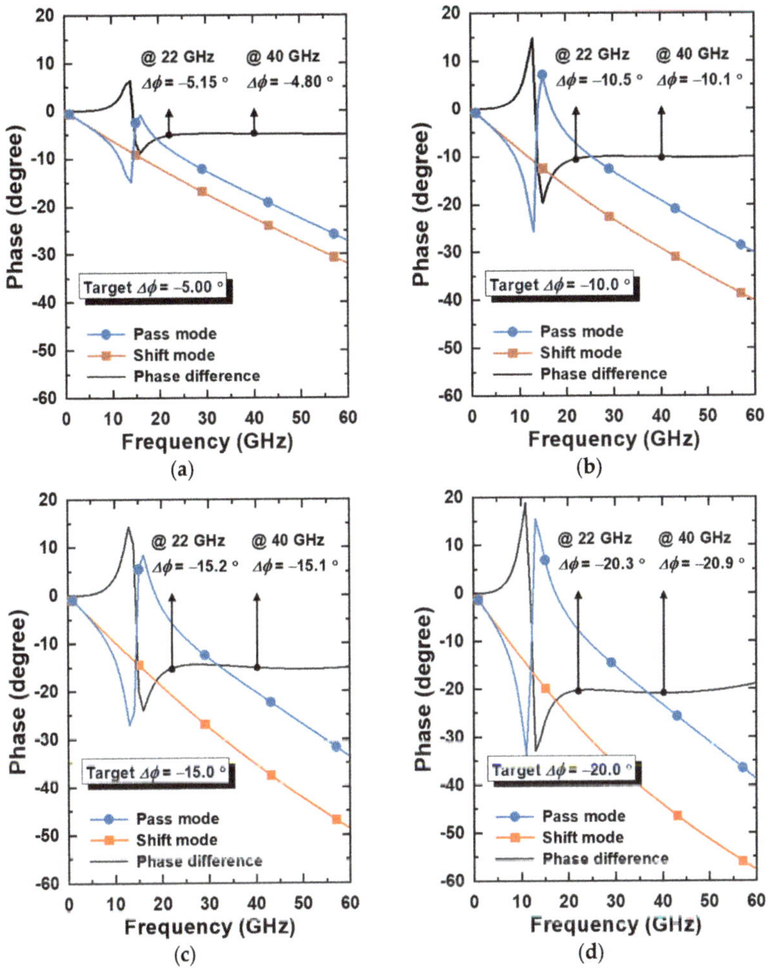

Figure 5. Simulation results of design examples of the wideband phase shifter: (**a**) 5°, (**b**) 10°, (**c**) 15°, and (**d**) 20°.

Table 1. Size of the used transistors, inductors, and capacitors.

Bits	5°	10°	15°	20°
M (μm) [1]	4.8	6.8	10.5	12.0
C_M (fF)	5.0	7.4	10.5	12.4
C_{SH} (fF)	29.1	47.6	80.5	106.2
L_{SH} (pH)	220	220	220	220
L_{RE} (nH)	1.73	1.44	1.44	1.13

[1] Total gate width.

Among the simulation results of the four designed phase shifters, the phase shifter of 20° was actually fabricated, measured, and analyzed. Accordingly, the phase shifter of 20° was investigated in more detail. As shown in Figure 5d, the simulation results of the phase difference of −20.9° were obtained at the 25 °C typical corner with the operating frequency of 40 GHz. In addition, simulation results at −40 °C, 25 °C, and 80 °C were examined in fast, typical, and slow corners to confirm the impact of process and temperature variations.

For the 20° phase shifter with the operating frequency of 40 GHz, simulation results of the phase difference of −19.2° and −22.3° were obtained in the −40 °C fast corner and 80 °C slow corner, respectively. It was confirmed as a simulation result that there was a deviation of approximately 3.1° by process and temperature variations with the operating frequency of 40 GHz.

As described above, in order to ensure wideband characteristics with suppressed phase errors, the slopes of the phase according to the frequency of shift and pass modes for each bit were designed to be the same. For all 5°-, 10°-, 15°-, and 20°-bits, the phase error from the operating frequency of 22 GHz to 40 GHz was within 0.6° under 25 °C typical corner conditions. If the operating frequency was expanded from 22 GHz to 60 GHz, the phase error was less than 0.97°. Such a frequency range includes all of the frequency ranges for 5G applications in the mm-Wave band. As a result, from the simulation results shown in Figure 5, it can be seen that the proposed design methodology is effective in securing wideband characteristics and suppressing phase errors in the phase shifter.

4. Measurement Results of the Designed Wideband Phase Shifter

In order to verify the effectiveness of the proposed design methodology through measurement, a phase shifter of the unit bit of the T-type L-C-L structure was designed using the 65-nm RFCMOS process. The phase shift target of the designed unit bit was set to 20°. For shift- and pass-modes, the gate voltages of the transistor were 1 V and 0 V, respectively. The measurement was carried out at room temperature. Figure 6 shows a chip photograph of the designed 20°-bit phase shifter. The chip and core sizes are 0.290 × 0.455 mm^2 and 0.130 × 0.260 mm^2. The designed phase shifter of the unit bit is actually integrated and used in the transceiver for the beamforming system. Therefore, although test pads were implemented in this study to verify the feasibility of the phase shifter itself, these test pads are removed when applied to the actual transceiver. On-wafer probes were used to measure RF input and output signals, and gate voltage was applied through a bonding wire.

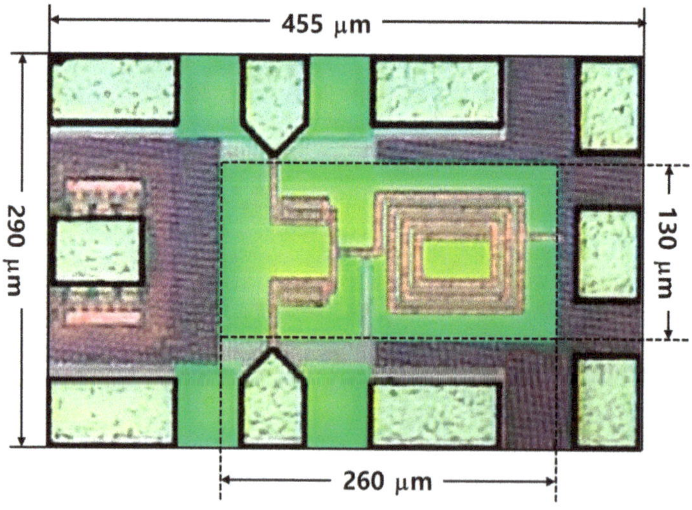

Figure 6. Chip photograph of the designed unit bit phase shifter.

Figure 5d shows the simulated phase characteristics according to the frequency of the designed unit bit phase shifter, and Figure 7 shows the measured phase characteristics. Due to the limitation of the measurement environment, the operating frequency of the designed unit bit phase shifter was measured up to 40 GHz. The target phase was 20°, but the measured phase was slightly reduced. One of the causes of the difference between measurement and simulation results may be the accuracy of EM simulation. In addition, as described in the previous section, process–voltage–temperature (PVT) variation can also be one of the important causes of the difference.

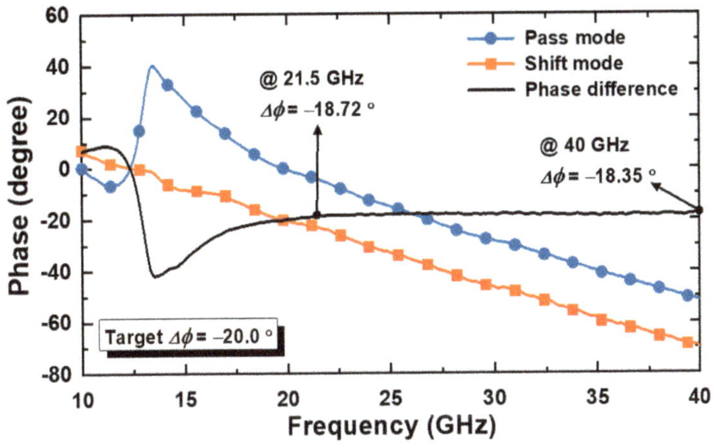

Figure 7. Measured phases for shift and pass modes and phase difference.

The measured phase differences of the shift and pass modes were 18.72° and 18.35° at 21.5 GHz and 40.0 GHz, respectively. In the operating frequency range of 21.5 GHz to 40.0 GHz, the measured phase difference was in the range of 17.85° to 18.72°. Therefore, based on the target phase difference of 20°, the phase error was less than 2.15°. However, considering the wideband characteristics of the proposed designed technique, the deviation of the phase difference in the operating frequency range from 21.5 GHz to 40.0 GHz was less than ±0.5°. Figure 8 shows the measured and simulated insertion losses of the designed unit bit phase shifter.

In Table 2, the performance of CMOS-based wideband phase shifters was compared. While phase shifters in most previous studies consist of several bits, the phase shift in this study consists of a unit bit. Therefore, accurate comparative evaluation is somewhat difficult. However, it can be seen that the phase shifter of this study generally has a low phase error in a wide frequency range.

Table 2. Performance comparison of CMOS phase shifters.

Ref.	Tech. (nm)	Type /Bits	Freq. (GHz)	BW (%)	IL [1] (dB)	RMS Phase Error (°)	P_{DC} (mW)	Core Size (mm^2)
[27]	28	VSPS/2.8° [2]	22–44	66.7	<5.81	<2.6	25	0.19
[28]	180	VSPS/22.5° [2]	27–33	20.0	<10.0	<4.0	6.6	0.44
[29]	65	RTPS/>360° [3]	27.8–31.2	11.5	<9.0	-	0	0.08
[30]	65	RTPS/>180° [3]	25–43	53.0	<9.1	-	0	0.15
[31]	65	STPS/5-bits	27–42	43.5	<14.5	<3.8 [4]	0	0.40
[32]	28	STPS/4-bits	29–37	24.2	<15.3	<8.8	0	0.07
This work	65	STPS/1-bits	21.5–40.0	63	<5.0	<0.5	0	0.03

[1] Insertion loss, [2] resolution, [3] phase-shift range, [4] controlled by bi-phase modulator.

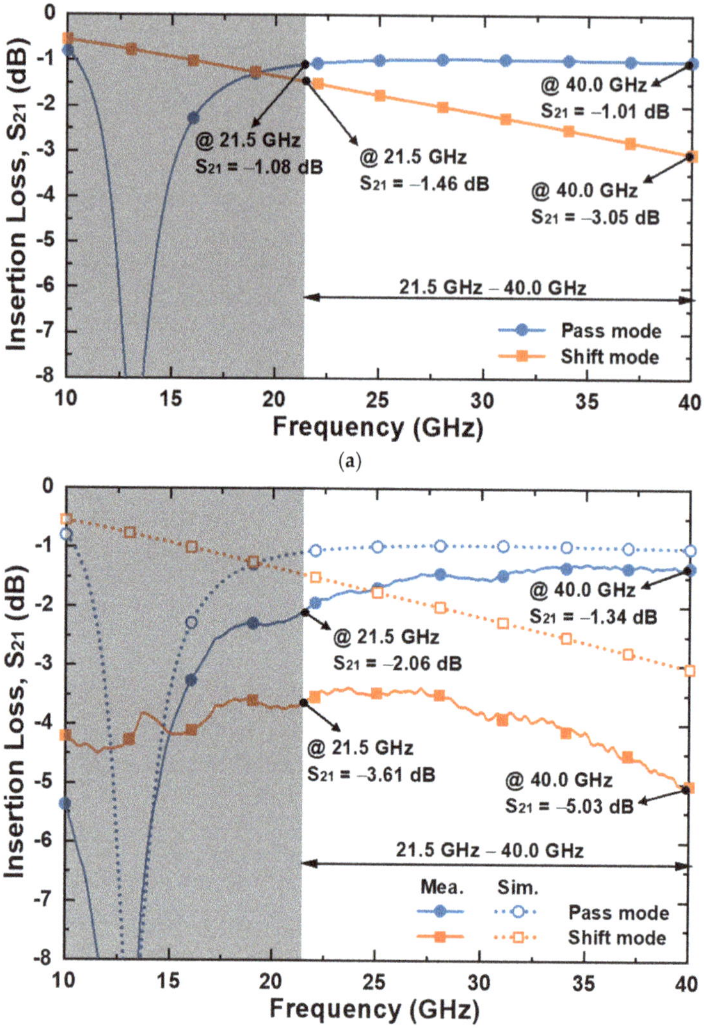

Figure 8. Insertion loss: (**a**) simulation and (**b**) measurement results.

5. Conclusions

In this study, we propose a design methodology for a wideband phase shifter with suppressed phase errors. To this end, the equivalent circuits in shift and pass modes of the phase shifter were analyzed. Through internal impedance analysis of equivalent circuits, it was shown using mathematical and simulation results that both modes can be represented by equivalent circuits of the L-C-L structure beyond a specific frequency. In shift and pass modes represented by equivalent circuits of L-C-L structure, it was confirmed that phase errors can be minimized in a wide range of frequency bands when the slope of the phase according to frequency is adjusted. Adjustment of the slope can be achieved by adjusting the frequency of -90° in each mode, which can be achieved by adjusting the inductance and capacitance in the equivalent circuit for each mode. The proposed design methodology was applied to a 20°-bit phase shifter designed with a 65 nm RFCMOS process. As a result of the measurement, at 21.5 GHz to 40.0 GHz, the phase error was within 0.87°.

Author Contributions: Conceptualization, S.J. and C.P.; methodology, S.J. and C.P.; investigation, S.J.; supervision, C.P.; writing—original draft, S.J.; review and editing C.P. All authors have read and agreed to the published version of the manuscript.

Funding: This work was supported in part by the National Research Foundation of Korea (NRF) through the Korean Government (MSIT) under Grant NRF-2021R1A2C1013666 and in part by the National Research Foundation of Korea (NRF) through the Korean Government (MSIT) under Grant NRF-2021R1A4A1032580.

Data Availability Statement: All the material conducted in the study is mentioned in the article.

Conflicts of Interest: The authors declare no conflict of interest.

References

1. Park, J.; Lee, S.; Chun, J.; Jeon, L.; Hong, S. A 28-GHz Four-Channel Beamforming Front-End IC with Dual-Vector Variable Gain Phase Shifters for 64-Element Phased Array Antenna Module. *IEEE J. Solid State Circuits* **2023**, *58*, 1142–1159. [CrossRef]
2. Mondal, S.; Carley, L.R.; Paramesh, J. Dual-Band, Two-Layer Millimeter-Wave Transceiver for Hybrid MIMO Systems. *IEEE J. Solid State Circuits* **2022**, *57*, 339–355. [CrossRef]
3. Al-Yasir, Y.I.A.; Abdulkhaleq, A.M.; Parchin, N.O.; Elfergani, I.T.; Rodriguez, J.; Noras, J.M.; Abd-Alhameed, R.A.; Rayit, A.; Qahwaji, R. Green and Highly Efficient MIMO Transceiver System for 5G Heterogenous Networks. *IEEE Trans. Green Commun. Netw.* **2022**, *6*, 500–511. [CrossRef]
4. Khan, B.; Tervo, N.; Jokinen, M.; Pärssinen, A.; Juntti, M. Statistical Digital Predistortion of 5G Millimeter-Wave RF Beamforming Transmitter under Random Amplitude Variations. *IEEE Trans. Microw. Theory Tech.* **2022**, *70*, 4284–4296. [CrossRef]
5. Chang, Y.-T.; Lin, K.-Y.; Wu, T.-L. Wideband Reconfigurable Power Divider/Combiner in 40-nm CMOS for 5G mmW Beamforming System. *IEEE Trans. Microw. Theory Tech.* **2022**, *70*, 1410–1422. [CrossRef]
6. Lee, S.; Hong, S. Frequency-Reconfigurable Dual-Band Low-Noise Amplifier with Interstage Gm-Boosting for Millimeter-Wave 5G Communication. *IEEE Microw. Wirel. Technol. Lett.* **2023**, *33*, 463–466. [CrossRef]
7. Lee, S.; Park, J.; Hong, S. Millimeter-Wave Multi-Band Reconfigurable Differential Power Divider for 5G Communication. *IEEE Trans. Microw. Theory Tech.* **2022**, *70*, 886–894. [CrossRef]
8. Capelli-Mouvand, T.; Deltimple, N.; Cathelin, P.; Ghiotto, A. Third-Order Intermodulations Phase Shifting in Power Amplifiers for Spatial Filtering in Multi-Beam 5G mmW Phased Array. *IEEE Trans. Circuits Syst. II Express Briefs* **2023**, *70*, 3363–3367. [CrossRef]
9. Wang, Z.; Wang, X.; Liu, Y. A Wideband Power Amplifier in 65 nm CMOS Covering 25.8 GHz–36.9 GHz by Staggering Tuned MCRs. *Electronics* **2023**, *12*, 3566. [CrossRef]
10. Chiou, H.-K.; Lin, H.-C.; Chang, D.-C. High-Efficiency and Cost-Effective 10 W Broadband Continuous Class-J Mode Quasi-MMIC Power Amplifier Design Utilizing 0.25 µm GaN/SiC and GaAs IPD Technology for 5G NR n77 and n78 Bands. *Electronics* **2023**, *12*, 3494. [CrossRef]
11. Ghaedi Bardeh, M.; Fu, J.; Naseh, N.; Paramesh, J.; Entesari, K. A Wideband Low RMS Phase/Gain Error mm-Wave Phase Shifter in 22-nm CMOS FDSOI. *IEEE Microw. Wirel. Technol. Lett.* **2023**, *33*, 739–742. [CrossRef]
12. Liao, Y.; Tang, M.; Pang, J.; Shirane, A.; Okada, K. A 19–34-GHz Bridged-T Phase Shifter with High-Pass Phase Compensation Achieving 3.9° RMS Phase Error for 5G NR. *IEEE Solid State Circuits Lett.* **2023**, *6*, 233–236. [CrossRef]
13. Singhal, N.; Hasan, S.M.R. A 25–30-GHz RMS Error-Minimized 360° Continuous Analog Phase Shifter Using Closed-Loop Self-Tuning I/Q Generator. *IEEE Trans. Very Large Scale Integr. (VLSI) Syst.* **2022**, *30*, 720–731. [CrossRef]
14. Park, G.-H.; Byeon, C.W.; Park, C.S. 60 GHz 7-Bit Passive Vector-Sum Phase Shifter with an X-Type Attenuator. *IEEE Trans. Circuits Syst. II Express Briefs* **2023**, *70*, 2355–2359. [CrossRef]
15. Akbar, F.; Mortazawi, A. An Integrated Compact Phase Shifter with a Single Analog Control. *IEEE Microw. Wirel. Technol. Lett.* **2022**, *32*, 410–413. [CrossRef]
16. Zhu, X.; Yang, T.; Chi, P.-L.; Xu, R. Novel Passive Vector-Sum Reconfigurable Filtering Phase Shifter with Continuous Phase-Control and Tunable Center Frequency. *IEEE Trans. Microw. Theory Tech.* **2022**, *70*, 1188–1197. [CrossRef]
17. Qiu, F.; Zhu, H.; Che, W.; Xue, Q. A K-Band Full 360° Phase Shifter Using Novel Non-Orthogonal Vector Summing Method. *IEEE J. Solid State Circuits* **2023**, *58*, 1299–1309. [CrossRef]
18. Dou, B.; Duan, Z.; Li, Y.; Ding, N.; Xiao, D.; Liao, B.; Yao, X. A 4–10 GHz Programmable CMOS Vector-Sum Phase Shifter for a Two-Channel Transmitter. *IEEE Trans. Circuits Syst. II Express Briefs* **2022**, *69*, 3699–3703. [CrossRef]
19. Qiu, F.; Zhu, H.; Che, W.; Xue, Q. A Simplified Vector-Sum Phase Shifter Topology with Low Noise Figure and High Voltage Gain. *IEEE Trans. Very Large Scale Integr. (VLSI) Syst.* **2022**, *30*, 966–974. [CrossRef]
20. Sung, E.-T.; So, C.; Hong, S. A 60-GHz Variable-Gain Phase Shifter with Particular-Sized Digital-RF Cells. *IEEE Trans. Microw. Theory Tech.* **2022**, *70*, 1302–1313. [CrossRef]
21. So, C.; Sung, E.-T.; Hong, S. A 60-GHz Variable-Gain Phase Shifter with an Active RL Poly-Phase Filter. *IEEE Trans. Microw. Theory Tech.* **2022**, *71*, 593–601. [CrossRef]

22. Li, X.; Liu, B.; Fu, H.; Ma, K. A 30–36 GHz Passive Hybrid Phase Shifter with a Transformer-Based High-Resolution Reflect-Type Phase Shifting Technique. *IEEE Trans. Circuits Syst. II Express Briefs* **2021**, *68*, 2419–2423. [CrossRef]
23. Meng, F.; Ma, K.; Yeo, K.S.; Xu, S. A 57-to-64-GHz 0.094-mm^2 5-bit Passive Phase Shifter in 65-nm CMOS. *IEEE Trans. Very Large Scale Integr. (VLSI) Syst.* **2016**, *24*, 1917–1925. [CrossRef]
24. Jang, S.; Kim, C.-Y.; Park, C. Design Technique of K-Band CMOS Phase Shifter with L-C-L T-Type Low Pass Structure. *Electronics* **2023**, *12*, 3678. [CrossRef]
25. Li, X.; Fu, H.; Ma, K.; Hu, J. A 2.4–4-GHz Wideband 7-Bit Phase Shifter with Low RMS Phase/Amplitude Error in 0.5-μm GaAs Technology. *IEEE Trans. Microw. Theory Tech.* **2022**, *70*, 1292–1301. [CrossRef]
26. Guan, P.; Jia, H.; Deng, W.; Dong, S.; Huang, X.; Wang, Z.; Chi, B. A 33.5–37.5-GHz Four-Element Phased-Array Transceiver Front-End with Hybrid Architecture Phase Shifters and Gain Controllers. *IEEE Trans. Microw. Theory Tech.* **2023**, *71*, 4129–4143. [CrossRef]
27. Zhou, J.; Qian, H.J.; Luo, X. High-Resolution Wideband Vector-Sum Digital Phase Shifter with On-Chip Phase Linearity Enhancement Technology. *IEEE Trans. Circuits Syst. I Regul. Pap.* **2021**, *68*, 2457–2469. [CrossRef]
28. Chang, Y.-T.; Ou, Z.-W.; Alsuraisry, H.; Sayed, A.; Lu, H.-C. A 28-GHz Low-Power Vector-Sum Phase Shifter Using Biphase Modulator and Current Reused Technique. *IEEE Microw. Wirel. Compon. Lett.* **2018**, *28*, 1014–1016. [CrossRef]
29. Gu, P.; Zhao, D. Ka-band CMOS 360° Reflective-Type Phase Shifter with ±0.2 dB Insertion Loss Variation Using Triple-Resonating Load and Dual-Voltage Control Techniques. In Proceedings of the IEEE Radio Frequency Integrated Circuits Symposium (RFIC), Philadelphia, PA, USA, 10–12 June 2018; pp. 140–143.
30. Lim, J.-T.; Choi, S.; Lee, E.-G.; Choi, H.-W.; Song, J.-H.; Kim, S.-H.; Kim, C.-Y. 25–40 GHz 180° Reflective-Type Phase Shifter using 65-nm CMOS Technology. In Proceedings of the European Microwave Conference (EuMC), Paris, France, 1–3 October 2019.
31. Tsai, J.-H.; Tung, Y.-L.; Lin, Y.-H. A 27–42-GHz Low Phase Error 5-Bit Passive Phase Shifter in 65-nm CMOS Technology. *IEEE Microw. Wirel. Compon. Lett.* **2020**, *30*, 900–903. [CrossRef]
32. Jung, M.; Min, B.-W. A Compact Ka-Band 4-bit Phase Shifter with Low Group Delay Deviation. *IEEE Microw. Wirel. Compon. Lett.* **2020**, *30*, 414–416. [CrossRef]

electronics

MDPI

Article

Millimeter-Wave Choke Ring Antenna with Broad HPBW and Low Cross-Polarization for 28 GHz Dosimetry Studies

Philip Ayiku Dzagbletey and Jae-Young Chung *

Department of Electrical and Information Engineering, Seoul National University of Science and Technology, Nowon-gu, Seoul 01811, Republic of Korea; philip@seoultech.ac.kr

* Correspondence: jychung@seoultech.ac.kr

Abstract: A choke ring horn antenna has been designed for use as an RF applicator in a compact range in vitro 28 GHz bioelectromagnetic exposure system. The 30 mm × 50 mm horn antenna was fabricated and measured to operate from 27.75 GHz to 34.5 GHz with a −20 dB measured S11 and a measured antenna gain of more than 10 dBi. A wide sectoral (flat top) and symmetric **E**- and **H**-plane pattern with a half-power beamwidth of more than 60 degrees was achieved with a cross-polarization discrimination of better than 28 dB. Electromagnetic slots were introduced in the antenna to suppress excess cavity mode radiation which inherently impacts the cross-polarization levels of choke ring antennas. The proposed antenna was successfully integrated into the compact measurement chamber in partnership with the Korea Telecommunication Research Institute (ETRI) and is currently in use for real-time 5G millimeter-wave dosimetry studies.

Keywords: bioelectromagnetics; choke ring horn antenna; cross-polarization; dosimetry; exposure system; horn antenna; measuring chamber; millimeter waves; 28 GHz; 5G

1. Introduction

In the wake of the call for increased data rates and system throughput, the millimeter-wave (mmWave) band has been of great focus for researchers in recent times. Although much of this frequency band from 30 GHz to 300 GHz is vastly untapped, there are many applications to which these bands are currently being utilized. These applications range from airport security scanning devices to IoT device-to-device and wearable gadget communications. Others include biomedical applications such as MRI and CT scans, mobile and broadband backhaul systems, military systems as well as the prospects for 5G mobile communication [1,2].

For the latter application, the 28 GHz band in particular, is a promising candidate for the current generation of IMT-2020 5G wireless and mobile communication standards. Most researchers consider 28 GHz a mmWave frequency band due to its close air interface resemblance characteristics to the mmWave band.

The migration to this unused frequency band for commercial purposes will, however, not be complete without the proper dosimetry of human cells. The most widely used methods for human tissue analysis are the in vivo and in vitro [3–5]. Unlike other commercially licensed frequency bands, there is much work to be done for both in vivo and in vitro dosimetry experiments in the 28 GHz frequency band. For the latter, the penetrative effects of the mmWave frequency on human tissues are studied with the cells placed on cultured plates or transwells inside a controlled radiation chamber.

An area of concern is the geometry of the chamber sizes at 28 GHz and how it influences the antenna and field properties. As will be seen in Section 2 of this communication, certain key performance indicators (KPIs) are used to establish the relationship between the field properties and antenna as well as the complete geometry of the chamber. It is worth noting that the short wavelength of the mmWave, coupled with strict chamber condition requirements, make the chamber geometry and the culture plate's properties and antenna performance very sensitive and critical [2,6].

Figure 1 shows a proposed exposure chamber for the intended experiment. A horn antenna or any suitable RF applicator with a certain electrical performance illuminates mmWave onto the test samples for a given time at a given exposure distance. In most mmWave in vitro propagation systems, 35 mm diameter Petri dishes are placed in the farfield region of the RF applicator, generating a plane wave as demonstrated in [7–9].

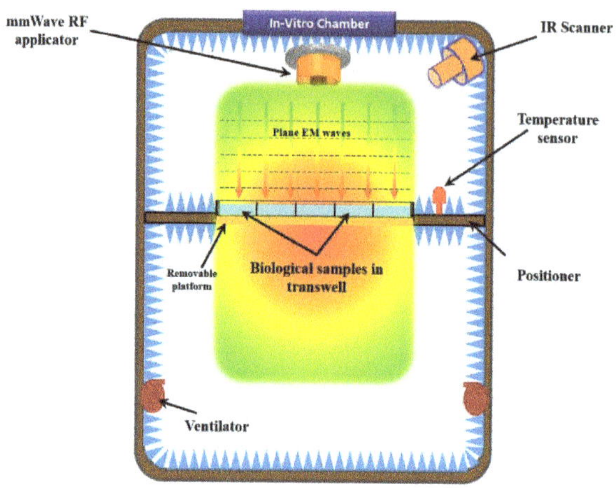

Figure 1. Proposed mmWave in vitro exposure system (not drawn to scale).

According to [7], most radiating systems do not use enclosures due to the complexity of controlling the environmental conditions inside the chamber. Nonetheless, in mmWave dosimetry, a proper enclosure system is required for the analysis of both thermal and ionizing experimentations [8]. Figure 1 thus shows a radiation box closing the antenna used for the EM wave emission. The enclosure mimics a typical incubator system or an anechoic chamber to trap fields within a controlled space.

The main challenge with the conventional radiating system approach is the use of a standard directive horn antenna which has an inherent drawback of poor exposure efficiency and also suffers from insufficient power density uniformity over the surface under test (SUT) [7,8]. This then requires a large test chamber with a small surface area; an option that is least desirable for commercial use. The best approach, as described by [8] and others regarding in vitro radiating and propagating dosimetry research at mmWave, is a modified horn antenna with broad beamwidth. To design a compact in vitro dosimetry chamber measuring 100 cm × 100 cm and operating at 28 GHz will require a specialized RF applicator, which is not available commercially. The compact chamber will require space for high frequency wave suppressants, ventilators, thermal cameras, positioners, etc., which then leave a compact space for the RF applicator. Thus, a 50 × 50 mm applicator is proposed. From the specified physical dimension of the radiator, more details on the values of Table 1 are elaborated in Section 2.1. A detailed chamber specification has been outlined in [10].

Table 1 shows a summary of the proposed RF applicator requirement for an in vitro mmWave exposure system. The antenna structure suitable to meet the criteria is usually a horn antenna with an oversized aperture and/or with several chokes. These structures provide a sectoral or secant pattern with a sharply descending slope. The choke ring antenna design is a known concept in antenna design [11–16]. The purpose of the technique is to alter the electric fields at the edges of a horn antenna to provide a secant beam (flat top) pattern.

Table 1. Invitro RF applicator specification.

Parameter	Specifications
Dimensions	$\leq$50 mm × 50 mm
Bandwidth	27.75~28.25 GHz
Gain	$\geq$8 dB
Sidelobe levels	≤-20 dB
Polarization	**E-H** symmetric polarization
HPBW	$\geq$60 degrees
Exposure efficiency	$\geq$60%
Spot size	$\geq$80 × 80 mm
S11	≤-20 dB

The CRA in [17], the smooth-walled horn antenna of [18] and the NASA X-band multiple CRA [12] may be suitable examples of antenna systems that can meet the stringent specifications in Table 1. These were both fabricated with 3D printing of a specific dielectric material and metalized in silver foam. The foam metallization with silver reduced the weight of the antenna, but more importantly suppressed unwanted higher cavity modes which are inherent to the all-metal design of such antennas.

To this end, an all-metal choke ring horn antenna has been proposed in this communication with excellent field properties for use as an RF applicator in a mmWave exposure system. A key feature of this design is the achievement of a highly sectoral beam pattern with a low-profile horn antenna, offering excellent cross-polarization to ensure a symmetric **E-H** pattern.

Section 2 outlines the bioelectromagnetic requirements of the chamber system and by extension, the antenna requirement. Section 3 introduces the choke ring horn antenna with a proposed technique to suppress the cross-polarization of the all-metal design. Subsequent sections show the measured performance of the proposed antenna for both impedance and radiation field performance, making it suitable for use as an RF applicator in the proposed chamber.

2. 5G Dosimetry Specifications

In the complete system design, certain key performance indicators (KPIs) will be considered to properly characterize the performance of the proposed mmWave exposure system. Such indicators may be categorized into three broad sections, namely the SUT, the field quantities and the RF applicator or antenna. SUT considerations will include the material geometry, be it a Petri dish or transwells, the material properties or the volume and size of the transwells.

Field properties indicative of the antenna performance within the confines of the chamber will include the **E**-field magnitude and phase, the power density and exposure efficiency. Others may include the illumination spot size and exposure distance. These quantities are influenced directly by the gain and radiation efficiency of the antenna, the beamwidth, cross-polarization and the **E-H** pattern in the farfield.

A detailed description of the compact exposure chamber has been presented in a previously published work by the same authors [10]. For brevity and focus on the antenna design, field properties of power density, exposure efficiency and illuminated spot size will be discussed.

2.1. RF Applicator/Antenna Specifications

With the given antenna specification in Table 1, the traditional horn antenna, as already mentioned, will not provide the wide HPBW to evenly distribute the required power to the surface area under test. The relationship between the directivity of a standard aperture antenna, beamwidth and height will limit the design of an antenna with 50 × 50 mm size to achieve a 60-degree HPBW and **E-H** symmetry. From [19], a broader HPBW and low gain horn antenna require a much shorter horn with a large flare angle. These two conditions are unrealizable at 28 GHz due to the small wavelength. An approach used by some researchers including [8,20] was to use dielectric loadings and choke insertions to obtain the required antenna properties.

Furthermore, the antenna must possess symmetric **E-H** plane patterns with low cross-polarization. These are essential for accurate power density and exposure uniformity calculations, as inference from neighboring polarization reduces the power density at a given exposure distance and hence the exposure efficiency.

2.2. Key Performance Indicators (KPI)

2.2.1. Power Density

The power density (P_D) is calculated by using the Poynting vector as described in Equation (1), where $\boldsymbol{E}$ and $\boldsymbol{H}$ are the electric and magnetic peak phasors, respectively. That is, the time-varying average energy transferred per unit area within the exposure enclosure of Figure 1, assuming a radiated power (*Prad*) of 1 W.

$$P_D = |0.5(E \times H^*)|_{Prad=1\text{W}} \tag{1}$$

References [9,21] as well as other researchers, use the specific absorption rate (SAR), which is specific to the cell type under test. The power density metric was used because the tissue sizes will be ordered by micrometer thicknesses and thus require a system performance that is independent of the samples under test.

The power density profile example as shown in Figure 2 measures the concentration of the radiated power or the field intensity at a predefined illumination spot size (*S*) on the SUT, which is usually at the center of the antenna. Ideally, the curve should be above the −0.5 dB or any specified uniformity level. In most in vitro mmWave dosimetry tests, −0.5 dB and −1.0 dB are specified [8], but they are subject to the system requirements. That is, a −0.5 dB uniformity translates to about 89% of the input power, incident on the SUT. It can, therefore, be deduced that larger distances can reach better uniformity levels with the standard horn antenna. This means a larger incubator is required when using a narrow beamwidth antenna. In this regard, the HPBW of the antenna will directly influence the field intensity on the SUT at a given distance.

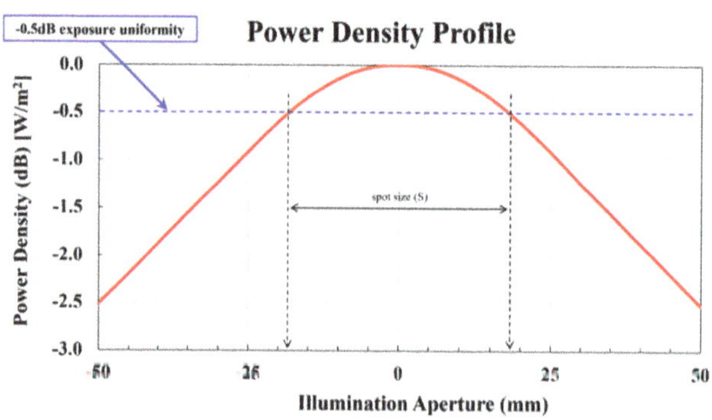

Figure 2. Power density profile example showing spot size.

The power density P_D in Equation (2) is calculated from the received power (*Prec*), where P_{rec} is the received power of the radiator in dBW. *Ae* is the antenna effective aperture, *G* is given as the antenna gain and λ is the wavelength in free space.

$$\begin{aligned} P_D &= {}^{\mathrm{P}_{rec}}/_{A_e} \\ A_e &= \lambda^2 G/4\pi \end{aligned} \tag{2}$$

The above equation holds for farfield received power [22]. In the nearfield, the value of P_{rec} is influenced by the nearfield path loss (P_{loss}) shown in Equation (3).

$$P_{loss(E)} = \frac{P_{rec(E)}}{P_{tx}} = \frac{G_{tx}G_{rec(E)}}{4} \cdot \left(\frac{1}{(kr)^2} - \frac{1}{(kr)^4} + \frac{1}{(kr)^6} \right) \quad (3)$$

For Equation (3), P_{tx} is the transmitted power of an open-ended probe used in verifying the RF applicator performance, r is the exposure distance between the antenna aperture opening and the surface of the SUT and k is represented by the wavenumber ($2\pi/\lambda$). Gtx is the gain of the open-ended probe and G_{rec} is the gain of the RF applicator. This loss term at higher frequencies is thus negligible and can thus be ignored if necessary.

2.2.2. Exposure Efficiency

The exposure efficiency is defined as the energy flux incident on the SUT. This is akin to the SAR since the chamber will be designed irrespective of the tissue used.

The energy flux is calculated with respect to the area of the plane under test. That is the area of a single dish or the surface area of any number of dishes. In the initial test, however, the rectangular area of the plane is used, giving an $L \times L$ sized SUT, as seen in Figure 3.

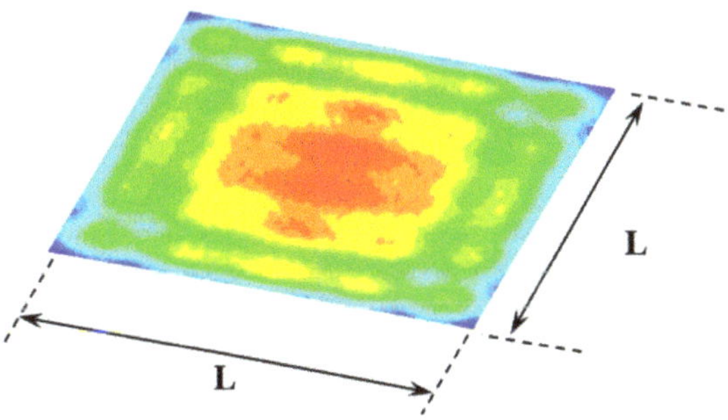

Figure 3. Power density curve on a plane.

The exposure efficiency is given by Equation (4), where P_{rad} is the total radiated power, S_m = time-averaged Poynting vector and A = surface area of the SUT calculated from Equation (5).

$$\eta_e = \frac{1}{P_{rad}} \iint_A S_m \delta A \quad (4)$$

$$A = \frac{\pi D^2}{4} \quad (5)$$

$$S = 2h * \tan\left(\frac{\varphi}{2}\right) \quad (6)$$

2.2.3. Illuminated Spot Size

This defines the area of uniform illumination on the SUT at a given uniformity. The optimal spot size for a given HPBW will be determined from Equation (6), where S is the spot size, h is the distance from the antenna aperture to the SUT and φ is the HPBW. At the given distance h, shown in Figure 4, a wider φ will achieve the same uniformity level but at a shorter exposure distance. Considering this, RF applicators with wide beamwidth are most desirable in building a compact exposure system. The value is obtainable from the power density profile curve. At a given uniformity level of −0.5 dB, a spot size of 26 mm is incident on the SUT for a given exposure distance and HPBW.

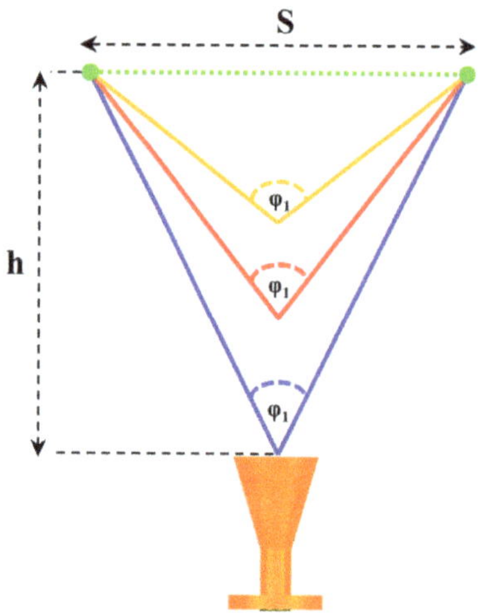

Figure 4. Spot size calculation.

These parameters will become important in the subsequent sections in determining the performance of the proposed horn antenna.

3. Antenna Design

With little research available for horn antenna design for use in an in vitro dosimetry at 28 GHz, it was necessary to remodel existing exposure system antennas, mostly greater than 28 GHz frequency bands to check their performance at 28 GHz. Four antenna models were chosen from several existing antennas at different millimeter-wave frequencies.

Figure 5 shows the models reoptimized at 28 GHz. The conical antenna model of Figure 5a without any dielectric loading is used for dosimetry analysis by [23–25] at 35 GHz/107 GHz, 60 GHz and 46 GHz, respectively. The pyramidal horn antenna shown in Figure 5b, which is commonly used for exposure system tests, were designed at 50 GHz, 60 GHz and 60.4 GHz by [9,26,27], respectively.

Ad hoc horn antenna models in Figure 5c,d were also modeled from [11,18] at 29.5 GHz and 60 GHz, respectively. The 29.5 GHz model in Figure 5d was designed for enhancing a directive beam pattern with a small form factor, but not for dosimetry analysis. As will be seen later, however, it proved capable after a few modifications to the dielectric contours.

A summary of the four antenna models is presented extensively with their corresponding E-field profile curves and power density profiles at 28 GHz. The conical horn has a nearly linear E-field magnitude curve with more of the power concentrated at the center of the antenna. The exposure uniformity and power density profiles show narrow spot sizes at the −0.5 dB uniformity level for different exposure distances (*h*).

The result of the pyramidal horn was similar to the conical one with an asymmetrical E-H plane pattern, but with a wider beamwidth compared to the latter. The exposure uniformity is also narrower at the center of the SUT. The choke ring antenna (CRA) model in Figure 5c has a more conformal r-squared E-field profile curve with a wide beamwidth of 50 degrees at an 11.5 dB gain. The exposure uniformity in Figure 5d is much larger due to the wider beamwidth. The E-field power, however, drops much faster in the farfield region. This indicates a limitation to the chamber size as it cannot be increased arbitrarily. The contoured smooth-walled model has a 55.5-degree beamwidth in both planes and has large uniformity values in the farfields above 4λg (guided wavelength).

Although not sufficient in illuminating the entire SUT at −0.5 dB exposure uniformity, the model in Figure 5c possessed a good potential to obtain the desired antenna properties in this research in that it has a second ring or choke to provide a secant farfield pattern for a broad and stable HPBW.

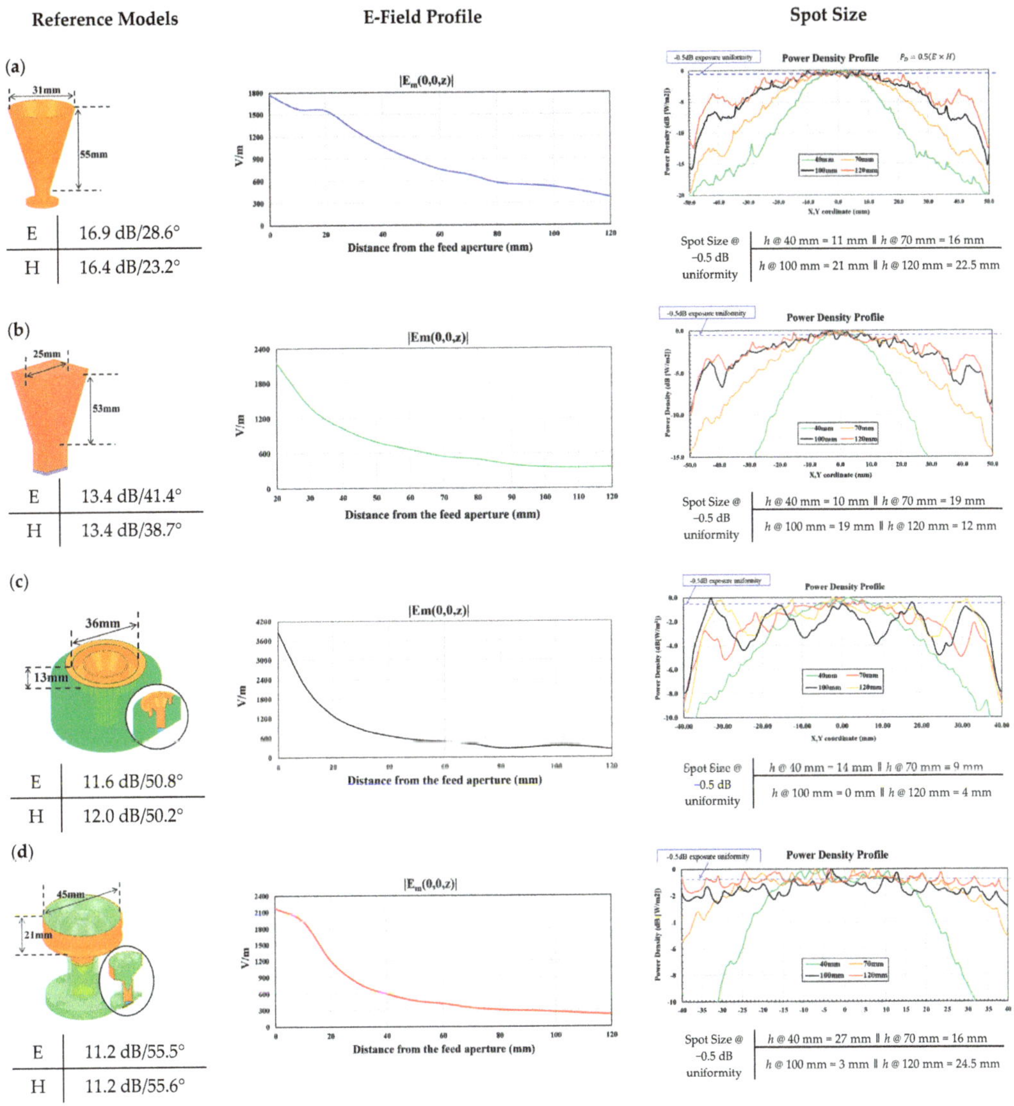

Figure 5. Remodeled horn antennas at 28 GHz with corresponding E-field and power density curves. (**a**) Conical Horn; (**b**) Pyramidal horn; (**c**) Choke ring horn; (**d**) contoured smooth-walled horn.

3.1. Choke Ring Horn Antenna (CRHA)

The concept of the secant beam shape is to design a theoretically "ill-fitted horn antenna" with a poor radiation pattern. That is, instead of a directive horn antenna, the dielectric and/or choke insertions create a secant beam in the farfield [8] thus making the HPBW broad with a nearly flat farfield radiation.

A rigorous analysis of the sectoral beam synthesis was performed in [28], but will not be covered in this communication. The chokes or extra rings in the horn antenna are designed to give rise to 180-degree phase shifts ($\lambda/2$) between each of the lobes with reference to the primary lobe generated by a traditional horn antenna. The depth of every successive choke is also chosen with the 180-degree phase shift. The beam coupling effect of the main beam and the successive chokes give rise to the sectoral beam. As the chokes increase, there is less coupling with the main beam and thus a reduced gain.

In the design analysis with Ansys software [29], the single choke ring naturally achieved high coupling with the center horn ring. At a 180-degree phase shift, the beam is out of phase with that of the main beam thus distorting the total beam formation and creating the secant beam pattern.

Figure 6 shows the initial design concept of the CRHA with ideal dimensions from [19]. The inner ring measured 14.75 mm at a flared opening of 31 degrees. The distance between the two rings was approximately $\lambda/2$, which provides a 180-degree phase shift. The 12.7 mm $\approx \lambda$ depth of the antenna also enabled the 180-degree phase shift of the choke to be realizable.

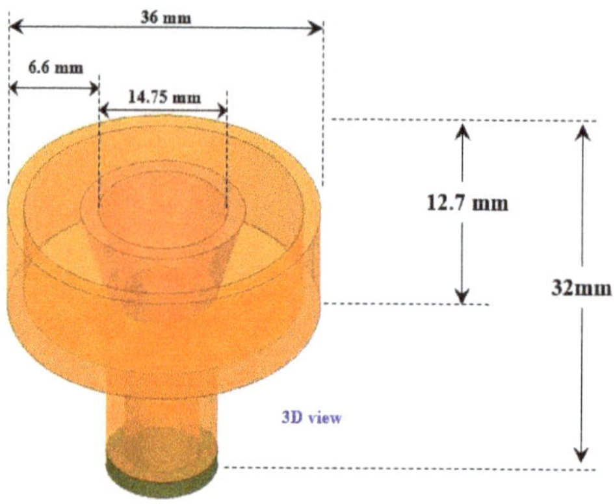

Figure 6. Design concept of CRHA with conical flared opening.

The E-field magnitude at a 150 mm exposure distance is also shown in Figure 7. The distribution is more uniform compared with [17,18]. However, the HPBW was observed at 54.4 degrees, which is less than the 60-degree requirement. Figure 7 shows the 3D radiation pattern using segmented FE-BI boundary condition in HFSS in which the radiating horn antenna is segmented from a 100 × 100 × 150 mm boundary box. The latter mimics an ideal boundary for the chamber system.

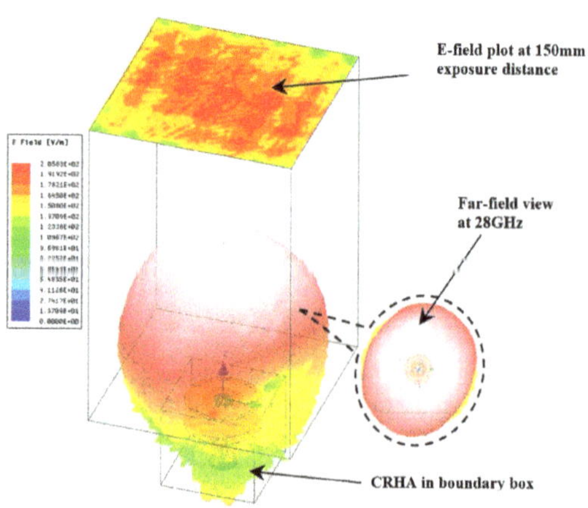

Figure 7. Simulated E-field and 3D radiation pattern of conical CRHA.

The inset farfield view at 28 GHz of the conical CRHA in Figure 7 shows an elliptical field distribution and a 45-degree tilt. A gain of 11 dB was realized with good **E-H** plane symmetry as shown in Figure 8. The 45-degree cut is also symmetric with a sharper secant slope. By minimizing the flare angle of the horn antenna and widening the choke ring, the elliptical field distribution was removed and a symmetric (**E-H**/45) farfield pattern is observed, as shown in the parameter sweep of Figure 9. From the sweep, the optimal symmetry was achieved at 8.76 mm $\approx$ WG-28 waveguide dimension (8.3 mm). The WG-28 waveguide will later be used as the coaxial to waveguide transition component. An optimal S11 of better than -35 dB is also achieved in Figure 10.

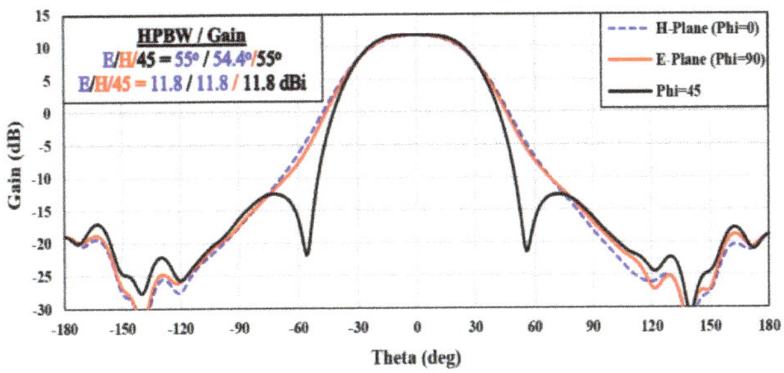

Figure 8. Simulated gain of conical CRHA at 28 GHz.

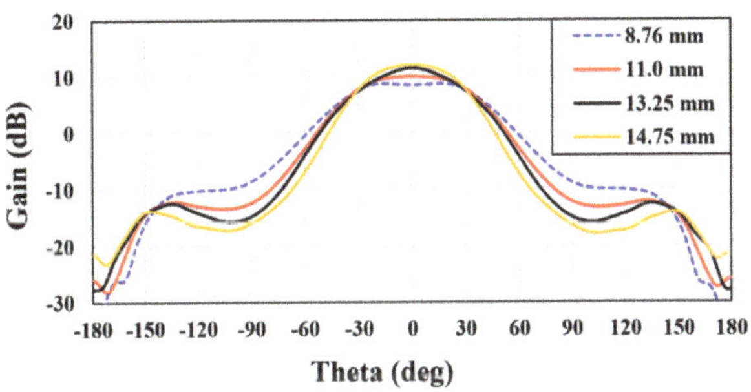

Figure 9. Parametric sweep for CRHA flare opening at 28 GHz.

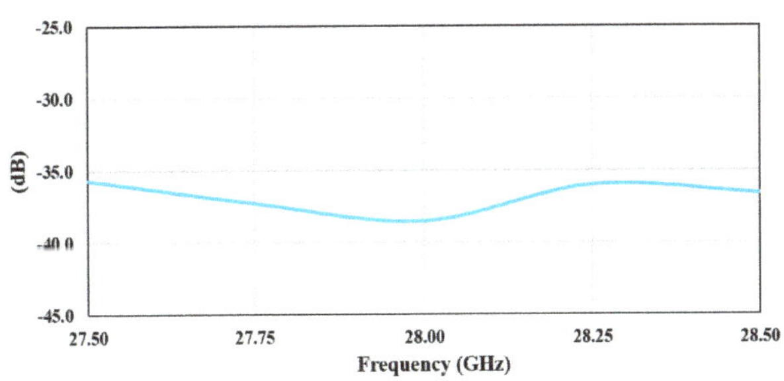

Figure 10. Simulated S11 curve of conical CRHA.

As the flared opening is reduced and approaches the dimensions of the center horn, the over-coupling between the center ring and the choke increases, thus increasing the secant angle and widening the beamwidth.

Although the cylindrical CRHA had a slightly reduced gain to 9.9 dB, as shown in Figure 11, the increased HPBW to 65.5 degrees was more desirable. This is an improvement from the conical CRHA and changes the conical choke horn into a cylindrical choke horn purposely for obtaining a wider beamwidth and secant pattern.

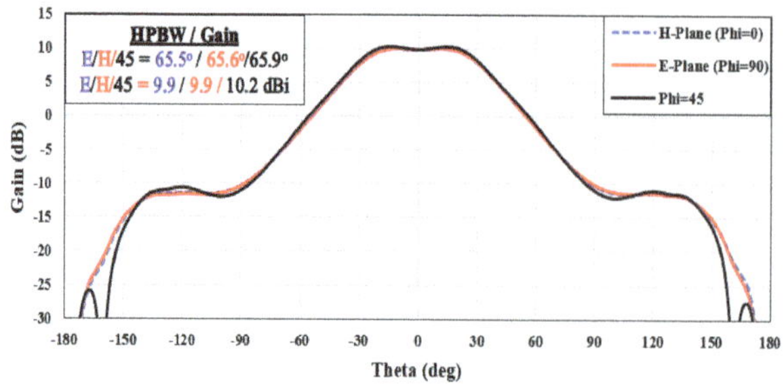

Figure 11. Simulated gain curve of cylindrical CRHA design concept.

Before fabrication, the cylindrical CRHA model is modified with a standard flange (50 mm flange diameter) and screws for ease of mounting. The best feeding method adopted was with the use of a circular to rectangular waveguide adapter operating from 21.7 to 33.0 GHz. In addition, a rectangular waveguide to coaxial transition is attached to the adapter to complete the feeding of the horn antenna. This latter transition also operated within 26.4 to 40.1 GHz band with a VSWR of better than 1.15.

Figure 12 shows the optimized CAD drawings of the cylindrical CRHA and flange. The stem of the antenna is reduced from 19.3 mm (Figure 6) to 10.2 mm (Figure 12); since the accompanying transition and adapter offered a wavelength longer than one at 28 GHz, which is necessary for the cavity TEM mode transition. This length reduction in the horn antennas does not reduce the performance, provided the transition distance is in multiples of λ at 28 GHz. The optimized model in Figure 13 includes an electromagnetic slot, which was introduced for the suppression of the cross-polarization.

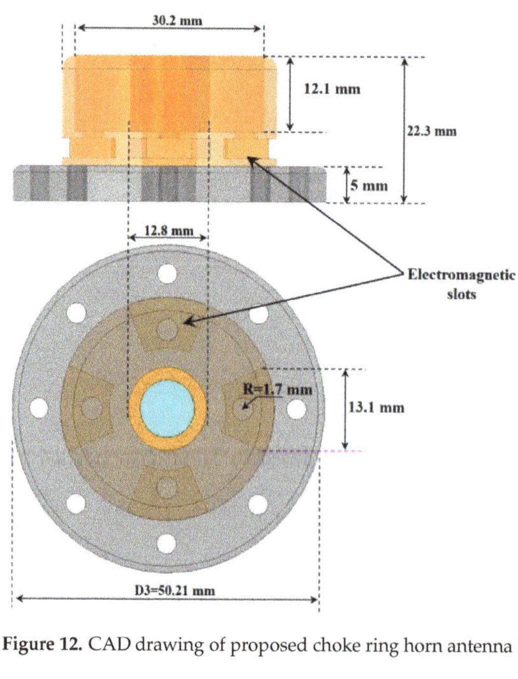

Figure 12. CAD drawing of proposed choke ring horn antenna (CRHA).

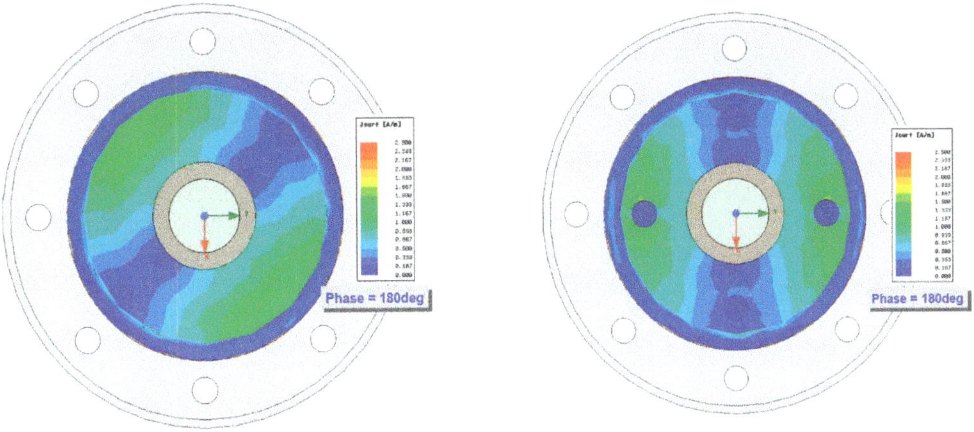

(**a**) (**b**)

Figure 13. Top view of CRHA showing surface current in the choke. (**a**) CRHA without EM slots. (**b**) CRHA with EM slots.

3.2. Cross-Polarization Suppression

The cross-polarization for most choke ring horn antennas is poor, as has been recorded by [11–15]. This is a result of the stronger coupling of the choke's surface currents with that of the center horn. Thus, it is a natural trade-off for the highly sectoral beam patterns. As such, a novel approach is introduced to suppress the cross-polarization of an all-metal choke ring horn antenna.

The design concept was inspired by observing the surface current distribution within the cylindrical CRHA as shown in Figure 13a,b. At a 180 degree phase, the choke had strong surface currents which created a highly sectoral radiation beam pattern. This, however, reduced the cross-polarization as higher-order modes are generated within the choke. As can be observed, there was a strong current distribution converging at the slots. The slots, therefore, provided an "escape" or alternate route to loosen the coupling within the choke, thus suppressing the cross-polarization. The loosening effect was due to the altering of the phase by the slots, which distorted the 180-degree phase shift required for a sectoral pattern. Therefore, the cross-polarization sectoral pattern was heavily distorted.

Figure 12 shows the CRHA with electromagnetic (EM) slots for cross-polarization suppression. This has four circular holes at the base of the choke. They are offset from the edge of the antenna and are 90 degrees apart. Another transverse slot was made at the base of the choke ring perpendicular to the circular slots and had a tapered aperture.

Figures 14 and 15 show the radiation patterns of the CRHA models with and without the transverse slots, respectively. The 4-transverse slot model achieved a cross-polarization reduction of more than 28.9 dB in the **E-H** plane and better than 10.5 dB in the 45-degree cut. Table 2 details the comparison of the CRHA without EM slots, with two slots and with four slots. As can be seen, there is minor change in the impedance values across the frequency band of interest. There is, however, a slight broadening of the HPBW with the slots compared with no slots, proving that the EM slots are an efficient and low-profile method of suppressing cross-polarization in the choke ring horn antenna. This, however, also results in marginal gain drops of about 0.5 dB. The cross-polarization discrimination (XPD) at the boresight also shows excellent results for both the 2-slot and 4-slot models.

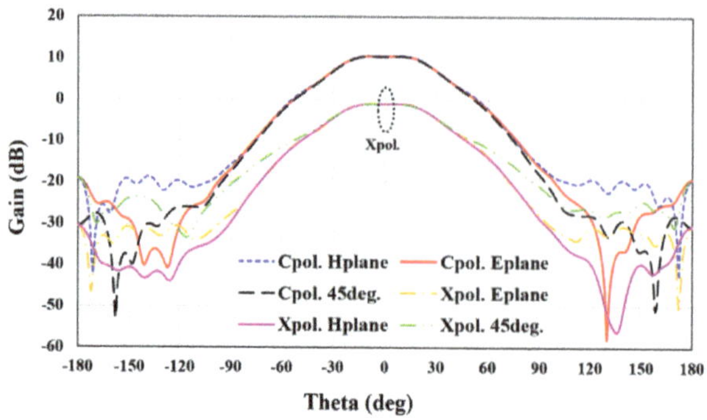

Figure 14. Simulated gain curve at 28 GHz of CRHA without EM slots.

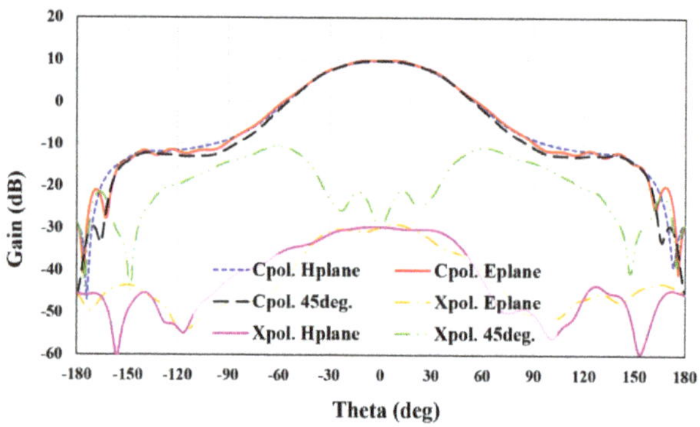

Figure 15. Simulated gain curve at 28 GHz of CRHA with EM slots.

Table 2. Cross-pol. suppression performance with EM Slots at 28 GHz.

Model	Direction	Cpol. (dB)	XPD (dB)	HPBW (deg.)	S11 (dB)
No slot	E-plane	9.9	11.5	65.5	<−32
	H-plane	9.9	11.6	65.6	
	45 deg.	10.2	11.6	65.9	
2-slots	E-plane	9.6	23.7	70.7	<−28
	H-plane	9.5	23.4	67.4	
	45 deg.	9.6	25.3	69.3	
4-slots	E-plane	9.8	38.7	67.6	<−27.5
	H-plane	9.8	39.4	66.3	
	45 deg.	9.8	38.3	67.2	

4. Fabrication and Measurement Discussion

The CAD model shown in Figure 12 was fabricated with braze (an alloy of copper and zinc) metal machining. The assembled antenna is shown in Figure 16. Braze material was used to give structural stability to the antenna and offer excellent radiation with anti-rust properties due to its copper and zinc content. Figure 16b also shows the cylindrical to rectangular waveguide adapter connected to the waveguide to coaxial transition mount used in the verification of the proposed radiator.

(**a**) (**b**)

Figure 16. Photos of fabricated and assembled CRHA. (**a**) CRHA without EM slots. (**b**) CRHA with EM slots.

As can also be observed in Figure 16a, the fabricated brass was polished with a sanding finish. Sanding is an industrial process of blasting the milled metal with a special abrasive material to smoothen the edges and remove surface contaminants. At high frequencies of 28 GHz, small imperfections in the milling can create unwanted modes, especially in the choke cavity. This process changes the polished color of the brass from bright gold to a dull but smooth finish.

The CRHA was placed in an anechoic chamber for farfield measurement as depicted in Figure 17a. The S-parameters were also measured in the setup shown in Figure 17b. Both measurements were performed at the millimeter-wave chamber facility of the Korea Radio Promotion Association (RAPA) facility in Yongsan, Seoul, Republic of Korea.

(**a**) (**b**)

Figure 17. Photos of fabricated and assembled CRHA. (**a**) Farfield measurement in anechoic chamber. (**b**) S-parameter measurement.

Figure 18 shows the measured S11 data from the setup in Figure 17b, including the simulation results. The measured S11 between 27.5 GHz to 28.5 GHz was better than −20 dB S11. The ripples in the measured results are a normal impedance response from the included attachments for measuring setup which were well-captured by the large frequency sampling of the network analyzer. The two attachments as shown in Figure 16 are the cylindrical to rectangular waveguide transition and the waveguide/coaxial adapter. Also, the mismatch between the simulation and measurement can be attributed to the loss factor from the aforementioned attachments, both of which were not considered in the simulation setup. The impedance results were nonetheless satisfactory.

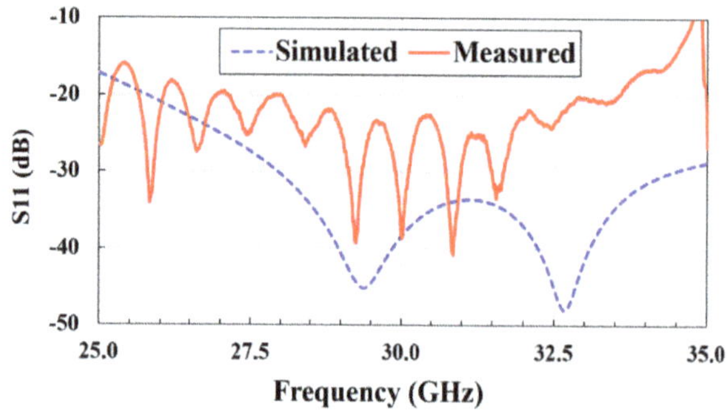

Figure 18. Simulated and measured S11 curves of proposed CRHA.

Figures 19–21 show the farfield measurement and simulation results for 27.75 GHz, 28 GHz and 28.25 GHz, respectively. The simulation results were optimized values from the final CAD model shown in Figure 12.

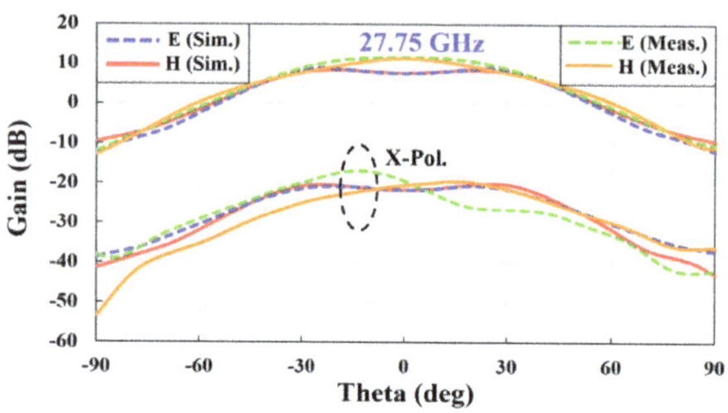

Figure 19. Simulated and measured gain of CRHA at 27.5 GHz.

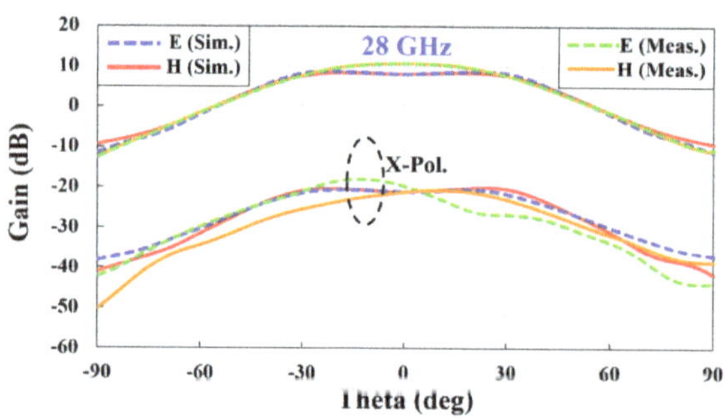

Figure 20. Simulated and measured gain of CRHA at 28 GHz.

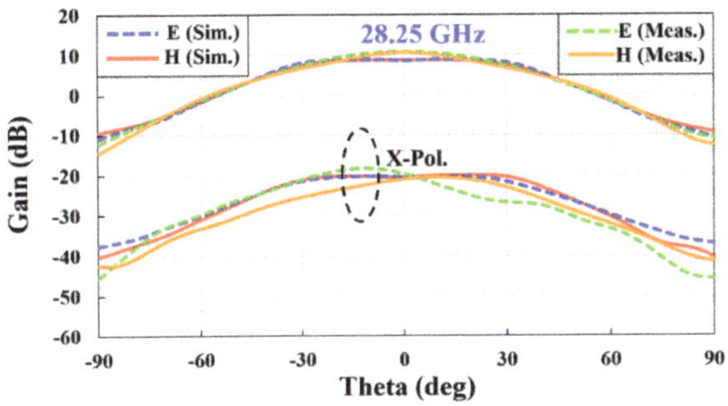

Figure 21. Simulated and measured gain of CRHA at 28.5 GHz.

At 28 GHz, the E-H plane was nearly symmetrical with a difference of about 1.4 degrees in HPBW, as seen in Figure 20. In all three frequencies, a 2.0 to 2.5 dB increase in gain is observed in the measured results compared to the simulation. This is theoretically about a 40% increase. A look at the radiation patterns shows conformal simulated and measured beam patterns, except for the increase in gain in the measured results.

As can be observed in Table 3, the increase in gain has a corresponding decrease in the HPBW for the measured results. HPBW remained at a 60-degree average across the measured frequencies. The measured cross-polarization discrimination (XPD) also remained consistent with the simulation.

Table 3. Cross-pol. Supression performace with EM slots at 27.75 GHz, 28 GHz, and 28.25 GHz.

Freq. (GHz)	Direction		Cpol. (dBi)	XPD (dB)	HPBW (deg.)
27.75	E-plane	Sim.	9.4	29.4	81.3
		Meas.	11.6	27.6	63.4
	H-plane	Sim.	9.9	29.5	79.9
		Meas.	11.6	30.4	50.0
28.0	E-plane	Sim.	9.5	28.5	79.3
		Meas.	10.8	28.4	59.9
	H-plane	Sim.	9.8	29.2	79.3
		Meas.	11.0	31.2	58.5
28.25	E-plane	Sim.	9.8	28.7	76.0
		Meas.	11.0	28.7	73.0
	H-plane	Sim.	9.0	29.2	77.3
		Meas.	11.0	30.8	58.8

5. Bioelectromagnetic Dosimetry Implementation

As already noted in the introduction, the accuracy of the exposure chamber hinges heavily on the proper radiation field synthesis during the antenna design simulation and in the measurement. Using the measurement setup shown in Figure 22, an open-ended waveguide probe is positioned at different distances (depicting the exposure distance) from the CRHA. At a specific exposure distance, z, the probe is panned in the xy direction to collect the received power from the CRHA.

Figure 22. Photo showing CRHA and open-ended waveguide probe in anechoic chamber.

From the received power, the E-field power and the power density were obtained and synthesized in Mathworks Matlab [30]. This case is true for both nearfield and farfield exposure distances. It must be noted, however, that both field calculations follow slightly different assumptions.

The measured power density profiles (PDPs) are plotted in Figure 23a–d. For each curve, two horizontal curves are shown for −0.5 dB and −1.0 dB uniformity. The exposure uniformity levels are used to determine how much power is emitted on the SUT per unit surface of a given distance. A −0.5 dB (0.89 W of 1 W input power) uniformity denotes the area on the SUT under which about 89% of the exposed output power is illuminated. The illuminated area will also influence the exposure efficiencies shown in Table 4.

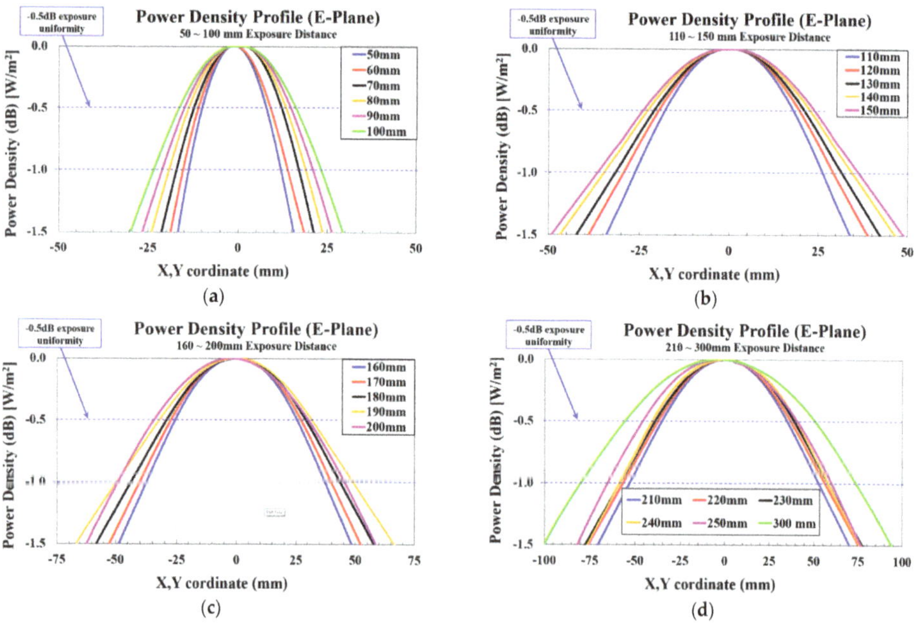

Figure 23. Measured E-field power density profiles with varying exposure distances. (**a**) 50~100 mm distance, (**b**) 110~150 mm distance, (**c**) 160~200 mm distance, (**d**) 210~300 mm distance.

Table 4. Measured field data summary at 28 GHz.

Exposure Distance (mm)	Spot Size −0.5 dB (mm)		Spot Size −1.0 dB (mm)		Peak Power Density/ob. Plane (mW/cm²)	Average Power Density/ob. Plane (mW/cm²)	Average Power Density/Spot Size (mW/cm²)		Z (Ω)	Peak E-Field (V/m)	Exposure Efficiency (%)	
	E	H	E	H			−0.5 dB	−1.0 dB			−0.5 dB	−1.0 dB
50	16	14	26	18	73.89	32.98	68.28	64.01	376.56	527.48	41.45	81.18
80	26	18	38	26	34.17	21.50	31.73	29.50	376.83	358.85	40.24	78.99
100	34	22	54	34	22.58	16.18	20.91	19.04	376.89	291.74	42.38	94.75
150	48	34	76	50	10.27	7.78	9.48	8.71	376.95	196.77	41.92	89.73
200	65	45	95	65	5.69	3.56	5.26	4.89	376.97	146.41	41.34	80.80
250	86	58	126	82	3.56	2.17	3.30	3.07	376.99	115.90	44.61	85.84
300	108	68	153	98	2.47	1.25	2.29	2.13	376.99	96.41	44.94	85.73

From the exposure efficiencies, it is observed that more of the radiated power is within the −1.0 dB uniformity area (about 80% of radiated power), which has more than 70% exposure efficiency. Close to 50% of the power is also within the −0.5 dB uniformity level.

Except for 27.75 GHz, the other frequencies (28.0 and 28.25 GHz) have average PDs greater than 2 mW/cm^2 for a 300 mm exposure distance. It can, therefore, be concluded that the average power density per spot size at 2.0 mW/cm^2 is likely to be beyond 300 mm distance. The 2 mW/cm^2 value is an acceptable level of power density exposure on human tissue. This value was also used as a reference by [8].

The E-field magnitude as seen in Figure 24a has a consistent r^2 asymptotic curve. This result offers insight into the plane wave property of the CRHA as having a linearly decaying phase from the nearfield into the farfield region.

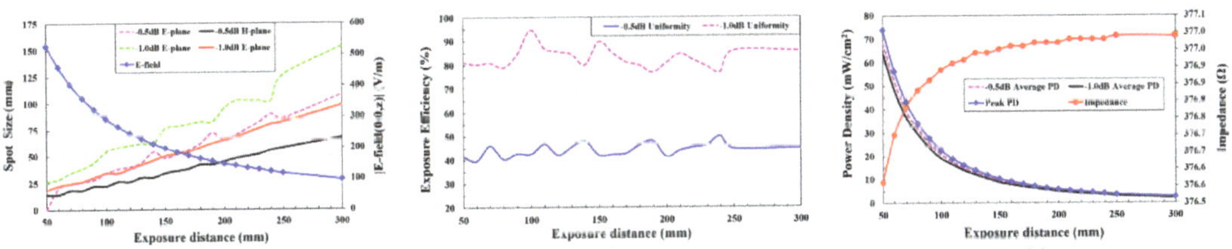

(a) (b) (c)

Figure 24. Measured and post-processed field properties of the proposed CRHA at 28 GHz. (**a**) Spot size and E-field magnitude. (**b**) Impedance and power density. (**c**) Exposure efficiency.

Also, from Figure 24a, the spot size is slightly asymmetric for both **E-H** planes. The slight dips in the values are because of insufficient exposure area in the measurement. This limitation is time-resource based, as one exposure distance measurement requires more than 3 h to collect sufficient data.

Figure 24b shows the exposure efficiency across the observable exposure distance. The slightly asymmetric **E-H** plane thus affected the exposure efficiency curves, especially for the −1.0 dB uniformity. A couple of fluctuations are also observed in the curve because of the insufficient exposure area. The average value of the exposure efficiency at −0.5 dB and −1.0 dB uniformity is consistent with that of [8].

The exposure efficiency reported in [8] is a simulated value and has an input power of 0 dB (1 W). With a 2-dB cable loss, the input power calculated was −2.0 dB (0.63 W). Thus, the efficiency calculations were lower than that of [8]. However, with a zero decibel input, the values of the measured exposure efficiency are higher than [8].

The peak power density (PD) and averaged power densities are also presented in Figure 24c. The results also show an r^2 decay, consistent with the E-magnitude curve in Figure 24a.

The impedance (**Z**) values in Figure 24c were recorded from the middle of the plane (0, 0, z). The impedance curve is also proportional to the exposure distance R. This confirms the assumptions made in [31–33] for the relationship between the impedance (*Z*) and the electric field magnitude (**E**) in obtaining the magnetic field density (**H**). That is, since an E-field open-ended probe was used in Figure 22, the H-field measured values were approximated from the E-field.

The first implementation of the proposed CRHA has been realized and is fully operational for dosimetry studies at the Electronics and Telecommunications Research Institute (ETRI), South Korea [10]. In this, a more detailed look into the compact 1 m × 1 m chamber is seen with the RF applicator shown in the inset photo overlooking a Petri dish holder. Under the Petri dish support is a probe and a positioner for *xyz* offsets. More details are shown in the communication from [10].

6. Conclusions

In this communication, a detailed outline of the design process for realizing a choke ring horn antenna have been presented for use as an RF applicator in a bioelectromagnetic dosimetry chamber. In line with the stringent electrical requirement to be met, the novel use of electromagnetic slots within the choke structure have been introduced to reduce the cross-polarization effect that is characteristic of metal-based choke ring horn antennas. This proposal achieved excellent results in both simulation and measurement, which have both been presented. Onward measurement and post-processing techniques have been presented to show the validity of the proposed RF applicator for dosimetry studies at 28 GHz.

Author Contributions: Conceptualization, P.A.D. and J.-Y.C.; methodology, P.A.D. and J.-Y.C.; software, P.A.D.; validation, P.A.D. and J.-Y.C.; formal analysis, P.A.D.; investigation, P.A.D.; resources, J.-Y.C.; data curation, P.A.D.; writing—original draft preparation, P.A.D.; writing—review and editing, P.A.D. and J.-Y.C.; visualization, P.A.D. and J.-Y.C.; supervision, J.-Y.C.; project administration, J.-Y.C.; funding acquisition, J.-Y.C. All authors have read and agreed to the published version of the manuscript.

Funding: This study was supported by the Research Program funded by SeoulTech (Seoul National University of Science and Technology).

Data Availability Statement: Data is contained within the article.

Conflicts of Interest: The authors declare no conflicts of interest.

References

1. Asri, M.; Tajdini, M.M.; Wig, E.; Rappaport, C.M. Automatic Threat Prediction of Body-Worn Objects for Security Screening Purposes. *IEEE Trans. Antennas Propag.* **2023**, *70*, 9732–9741. [CrossRef]
2. Accardo, J.; Chaudhry, A.M. Radiation exposure and privacy concerns surrounding full-body scanners in airports. *J. Radiat. Res. Appl. Sci.* **2014**, *7*, 198–200. [CrossRef]
3. Patrignoni, L.; Hurtier, A.; Orlacchio, R.; Joushomme, A.; Gannes, F.P.D.; Lévêque, P.; Arnaud-Cormos, D.; Revzani, H.R.; Mahfouf, W.; Garenne, A.; et al. Evaluation of mitochondrial stress following ultraviolet radiation and 5G radiofrequency field exposure in human skin cells. *Bioelectromagnetics* **2024**, *45*, 110–129. [CrossRef] [PubMed]
4. Morelli, M.S.; Gallucci, S.; Siervo, B.; Hartwig, V. Numerical Analysis of Electromagnetic Field Exposure from 5G Mobile Communications at 28 GHZ in Adults and Children Users for Real World Exposure Scenarios. *Int. J. Environ. Res. Public Health* **2021**, *18*, 1073. [CrossRef] [PubMed]
5. Ijima, E.; Kodera, S.; Hirata, A.; Hikage, T.; Matsumoto, A.; Ishitake, T.; Masuda, H. Excessive whole-body exposure to 28 GHz quasi-millimeter wave induces thermoregulation accompanied by a change in skin blood flow proportion in rats. *Front. Public Health* **2023**, *11*, 1225896. [CrossRef] [PubMed]
6. Orlacchio, R.; Andrieu, G.; Joushomme, A.; Patrignoni, L.; Hurtier, A.; Gannes, F.P.D.; Lagroye, I.; Percherancier, Y.; Arnaud-Cormos, D.; Leveque, P. A Novel Reverberation Chamber for In Vitro Bioelectromagnetic Experiments at 3.5 GHz. *IEEE Trans. Electromagn. Compat.* **2023**, *65*, 39–50. [CrossRef]
7. Paffi; Apollonio, F.; Lovisolo, G.A.; Marino, C.; Pinto, R.; Repacholi, M.; Liberti, M. Considerations for Developing an RF Exposure System: A Review for in vitro Biological Experiments. *IEEE Trans. Microw. Theory Tech.* **2010**, *58*, 2702–2714. [CrossRef]
8. Boriskin, A.; Zhadobov, M.; Steshenko, S.; LeDréan, Y.; LeCoq, L.; Person, C.; Sauleau, R. Enhancing Exposure Efficiency and Uniformity Using a Choke Ring Antenna: Application to Bioelectromagnetic Studies at 60 GHz. *IEEE Trans. Microw. Theory Tech.* **2013**, *61*, 2005–2014. [CrossRef]

9. Zhao, J.X. Numerical dosimetry for cells under millimetre-wave irradiation using Petri dish exposure setups. *Phys. Med. Biol.* **2005**, *50*, 3405–3421. [CrossRef] [PubMed]
10. Lee, Y.S.; Dzagbletey, P.A.; Chung, J.-Y.; Jeon, S.B.; Lee, A.-K.; Kim, N.; Song, S.J.; Choi, H.-D. Implementation of an in vitro exposure system for 28 GHz. *ETRI J.* **2020**, *42*, 837–845. [CrossRef]
11. Boriskin, A.; Zhadobov, M.; Person, C.; Sauleau, R. Focusing choke ring antenna for a short-range millimeter-wave exposure system. In Proceedings of the European Microwave Conference (EuMC), Nuremberg, Germany, 6–10 October 2013.
12. Chahat, N.; Amaro, L.R.; Harrell, J.; Wang, C.; Estabrook, P.; Butman, S.A. X-Band Choke Ring Horn Telecom Antenna for Interference Mitigation on NASA's SWOT Mission. *IEEE Trans. Antennas Propag.* **2016**, *64*, 2075–2082. [CrossRef]
13. Garcia-Muller, P.L. Optimisation of compact horn with broad sectoral radiation pattern. *Electron. Lett.* **2001**, *37*, 337–338. [CrossRef]
14. Jeon, K.-J.; Lee, K.; Son, J.; Lee, T.-K.; Lee, J.-W.; Lee, W.-K. X-band isoflux pattern antenna for SAR data transmission. In Proceedings of the 2011 3rd International Asia-Pacific Conference on Synthetic Aperture Radar (APSAR), Seoul, Republic of Korea, 26–30 September 2011.
15. Vacchione, J.D.; Kruid, R.C.; Prata, A.; Amaro, L.M.; Mittskus, A.P. Telecommunications antennas for the Juno Mission to Jupiter. In Proceedings of the 2012 IEEE Aerospace Conference, Big Sky, MT, USA, 3–10 March 2012.
16. Sun, M.; Qian, J.; Bai, X.; Zhou, L.; Lv, Y.; Kong, W.Z.F.; Jin, R. Compact Hybrid Choke Rings for Dual-Band Circularly Polarized GPS Antenna. *IEEE Antennas Wirel. Propag. Lett.* **2023**, *22*, 9–13. [CrossRef]
17. Boriskin, V.; Zhadobov, M.; Diedhiou, D.; Sauleau, R. Advanced feed for a 60-GHz exposure chamber. In Proceedings of the The 8th European Conference on Antennas and Propagation (EuCAP 2014), The Hague, The Netherlands, 6–11 April 2014.
18. Rolland, A.; Nguyen, N.T.; Sauleau, R.; Person, C.; Coq, L.L. Smooth-Walled Light-Weight Ka-Band Shaped Horn Antennas in Metallized Foam. *IEEE Trans. Antennas Propag.* **2012**, *60*, 1245–1251. [CrossRef]
19. Balanis, A. *Antenna Theory: Analysis and Design*, 4th ed.; John Wiley & Sons, Inc.: Hoboken, NJ, USA, 2016.
20. Kouzai, M.; Nishikata, A.; Saka, T.; Watanabe, S. Characterization of 60GHz Millimeter-Wave Living Experiments. In Proceedings of the IEICE EMC Conference, Kyoto, Japan, 20–24 July 2009.
21. Zhao, J.; Lu, H. Reduction of Exposure Inhomogeneity for Millimeter-Wave Experiments on Cells In Vitro. *IEEE Trans. Microw. Theory Tech.* **2015**, *63*, 533–545. [CrossRef]
22. IEEE Standards Working Group for Wireless Personal Area Networks (WPANs). *Near Field Channel Model. IEEE P802.15 Wireless Personal Area Networks*; IEEE: New York, NY, USA, 2004.
23. Rosenthal, S.W.; Birenbaum, L.; Kaplan, I.T.; Metlay, W.; Snyder, W.Z.; Zaret, M.M. Effects of 35 and 107 GHz CW Microwaves on the Rabbit Eye. In Proceedings of the USNC/URSI Annual Meeting, Boulder, CO, USA, 20–23 October 1976.
24. Sasaki, K.; Sakai, T.; Nagaoka, T.; Wake, K.; Watanabe, S.; Kojima, M.; Hasanova, N.; Sasaki, H.; Sasaki, K.; Suzuki, Y.; et al. Dosimetry Using a localized exposure system in the millimeter-wave band for in vivo studies on ocular effects. *IEEE Trans. Microw. Theory Tech.* **2014**, *62*, 1554–1564. [CrossRef]
25. Zhao, J.; Wei, Z. Numerical modeling and dosimetry of the 35 mm Petri dish under 46 GHz millimeter wave exposure. *Bioelectromagnetics* **2005**, *26*, 481–488. [CrossRef] [PubMed]
26. Zhadobov, M.; Nicolaz, C.N.; Sauleau, R.; Desmots, F.; Thouroude, D.; Michel, D.; Dréan, Y.L. Evaluation of the Potential Biological Effects of the 60-GHz Millimeter Waves Upon Human Cells. *IEEE Trans. Antennas Propag.* **2009**, *57*, 2949–2956. [CrossRef]
27. Quément, L.; Nicolaz, C.N.; Zhadobov, M.; Desmots, F.; Sauleau, R.; Aubry, M.; Michel, D.; Dréan, Y.L. Whole-genome expression analysis in primary human keratinocyte cell cultures exposed to 60 GHz radiation. *Bioelectromagnetics* **2012**, *33*, 147–158. [CrossRef] [PubMed]
28. Brachat, P. Sectoral Pattern Synthesis with Primary Feeds. *IEEE Trans. Antennas Propag.* **1994**, *42*, 484–491. [CrossRef]
29. ANSYS® Electronics Desktop. High Frequency Structure Simulator (HFSS). Available online: https://www.ansys.com/products/electronics/ansys-hfss (accessed on 23 July 2024).
30. *Matlab*, R2023b; Mathworks Inc.: Natick, MA, USA, 2023.
31. Capps, C. Delphi Automotive Systems. Near Field or Far Field? 16 August 2001. Available online: https://www.edn.com (accessed on 22 July 2018).
32. MICRONIX Corp. Near Field and Far Field. Available online: https://micronix-jp.com/english/tech/technology/emc_eng.html (accessed on 5 June 2018).
33. Armstrong, K. Guide to Testing Conducted Emissions. The Electrical and Electronics Engineers Guide to Design, Testing and Compliance, 1 July 2011. Available online: https://incompliancemag.com/article/guide-to-testing-conducted-emissions-based-on-the-methods-in-en-55022-and-en-55011/ (accessed on 10 March 2024).

electronics

MDPI

Communication

Time Domain Simulated Characterization of the Coplanar Waveguide in an On-Chip System for Millimeter Waveform Metrology

Kejia Zhao [1,*], He Chen [1], Xiangjun Li [2,*], Jie Sun [3], Bo Li [1], Dexian Yan [2] and Lanlan Li [1]

[1] National Institute of Metrology, Beijing 100029, China; chenhe@nim.ac.cn (H.C.); libo@nim.ac.cn (B.L.); lill@nim.ac.cn (L.L.)
[2] Center for THz Research, China Jiliang University, Hangzhou 310018, China; yandexian1991@cjlu.edu.cn
[3] Zhejiang Institute of Metrology, Hangzhou 310018, China; jiesun1022@163.com
* Correspondence: zhaokj@nim.ac.cn (K.Z.); xiangjun_li@cjlu.edu.cn (X.L.)

Abstract: We investigate the time domain characterization of a coplanar waveguide (CPW) based on an on-chip electro-optic sampling (EOS) system for millimeter waveform metrology. The CPW is fabricated on a thin layer of low-temperature gallium arsenide (LT-GaAs), and the substrate material is GaAs. A femtosecond laser generates and detects ultrashort pulses on the CPW. The forward propagating pulses are simulated using a simplified current source for the femtosecond laser at different positions on the CPW for the first time. Then, the influences of the CPW geometry parameters on the measured pulses are discussed. The varying slot width has larger influences on the amplitude of millimeter wave pulses than the center conductor width and the pumping gap. Finally, in the frequency range of 10 GHz to 500 GHz, the transfer functions calculated by the time domain pulses are in good agreement with the transfer functions calculated by the frequency domain ports. The above results are important for improving the measurement precision of the millimeter waveform on the CPW for millimeter waveform metrology.

Keywords: coplanar waveguide; electro-optic sampling; millimeter wave; waveform metrology

1. Introduction

Currently, the millimeter wave (60 GHz~300 GHz) is increasingly utilized in several key applications, such as high-speed communications, non-destructive sensing, spectroscopy, and imaging [1–5]. Ultrafast switching and extremely high-frequency operation are key issues for these applications. However, there are great challenges in the uniform measurements of ultrashort pulse generators in the millimeter wave range based on electronic components in ultra-wideband oscilloscopes and related technology [6]. Fortunately, an alternative approach was found based on the electro-optic sampling (EOS) principle compared to traditional sampling oscilloscopes with a bandwidth smaller than 50 GHz [7,8]. Using femtosecond laser pulses as the sampling pulses, EOS technology can extend its working bandwidths into the terahertz range. It allows the simultaneous measurement of the amplitude and phase of the millimeter wave or the THz pulse electric field in the time domain, with subsequent Fourier transform providing spectral characteristics in the 0.1~30 THz frequency range [9]. EOS for measuring the parameters of ultrashort electrical pulses has been verified as a promising method of full waveform metrology since 2001 [10]. In 2018~2019, the first international comparisons of the initial EOS apparatus operated by the National Institute of Standards and Technology (NIST), the Physikalisch-Technische Bundesanstalt (PTB), and the National Institute of Metrology (NIM) were conducted using a bandwidth of more than 110 GHz [11]. At present, pulses with a bandwidth of more than 10 THz can be measured using the EOS system [12–16].

The core part of the popular EOS system for millimeter waveform metrology is a combination of a coplanar waveguide (CPW) and a photoconductive structure [8,17]. The time domain measurement of the voltage pulses can be carried out at different positions along the waveguide, which makes it possible to separate voltage pulses propagating in forward and backward directions. By using digital signal-processing algorithms, those temporally overlapped waveforms will be divided completely. This separation method is equivalent to using directional flexible couplers in VNAs, which constitutes the main advances of the EOS system in millimeter waveform metrology. At present, the time domain measurement is able to generate an epoch of over 2 ns corresponding to a frequency spacing of 500 MHz, and the usable bandwidth of the voltage pulses extends up to 500 GHz [17]. Another advantage of this configuration is that the same laser source is applied to generate and detect the ultrashort voltage pulses in the same substrate, which significantly simplifies the experimental setup and reduces the uncertainty of electro-optic (EO) measurements. However, this kind of EOS measurement demands good characterization of the on-wafer coplanar waveguide and de-embedding of the probe [6–8]. Additionally, the voltage pulses must be transferred from the CPW to a coaxial waveguide of the device under test. Though there are many experimental results [8,17], the propagation mechanism of pulses along the CPW is still unclear and should be further investigated by full electromagnetic wave simulation tools, such as the Finite Element Method (FEM) or the Finite Difference Time domain (FDTD).

In this paper, we investigate the time domain characterization of the CPW based on an on-chip EOS system for millimeter waveform metrology. However, if we simulate the whole generating and detecting procedure of the pulse waveform, including optical, millimeter electromagnetic waves and the time-dependent carrier dynamics, in the semiconductor simulation, the computational cost is usually huge and needs a very long time [18]. To improve the research efficiency, the millimeter wave pulse generation process can be simplified by introducing a current source that is equivalent to the laser pulse excitation [19]. Here, we characterize the CPW using both the time domain and the frequency domain simulations. The CPW is fabricated on a thin layer of low-temperature gallium arsenide (LT-GaAs) using a GaAs substrate. For the first time, we simulate the forward propagating pulses at different positions on the CPW by simplifying the excitation of a laser pulse as a current source, which enables the highly efficient characterization of the EOS system before and after the experiments. Meanwhile, the influence of several CPW geometry parameters on the measured pulses is well discussed by characteristic impedance in the circuit theory and fast Fourier translation. Finally, the transfer functions of the CPW with different lengths are calculated by the pulses at different positions. There is good consistency between the results from a transformation of the time domain pulses and direct simulations from the frequency domain. The undergoing propagation rules of pulses along the CPW are useful for the future realization of EOS for millimeter waveform metrology.

2. Design Scheme and Research Method

The on-chip system based on EOS technology for millimeter waveform metrology is schematically demonstrated in Figure 1. The ultrashort pulses are produced by focusing the femtosecond laser beam with an 800 nm center wavelength (referred to as the pump beam) onto a biased photoconductive gap on the center conductor of a 2-mm-long CPW. The full width at half maximum (FWHM) of the optical pulses is about 100 fs as generated from an autocorrelation. Figure 1 shows the conventional CPW structure with signal and ground lines on a GaAs substrate. The fundamental dimensional parameters are the real permittivity of GaAs $\varepsilon_{GaAs} = 12.7$, the signal width $w = 45$ μm, the air slot width $s = 30$ μm, the gold thickness $t_m = 0.5$ μm, the LT-GaAs thickness $t_{d1} = 1$ μm, and the GaAs thickness $t_{d2} = 200$ μm. For full-wave electromagnetic (EM) analysis, the structure is enclosed in a box with perfect match layer (PML) boundaries. At the left edge of the CPW, a photoconductive structure as a gap $g = 10$ μm is integrated into the center conductor, which is biased by a 25 V voltage source. The waveguide has been fabricated on a 1-micrometer-thick layer of LT-GaAs with a shorter carrier lifetime, causing the optically induced conductivity to quickly vanish in the photoconductive gap. The main advantage of the structure is that one laser source can completely generate and detect ultrashort pulses in the same material without external EO probes [20]. This on-chip system significantly simplifies the experimental setup and reduces the uncertainty of EOS measurements for millimeter waveform metrology [6].

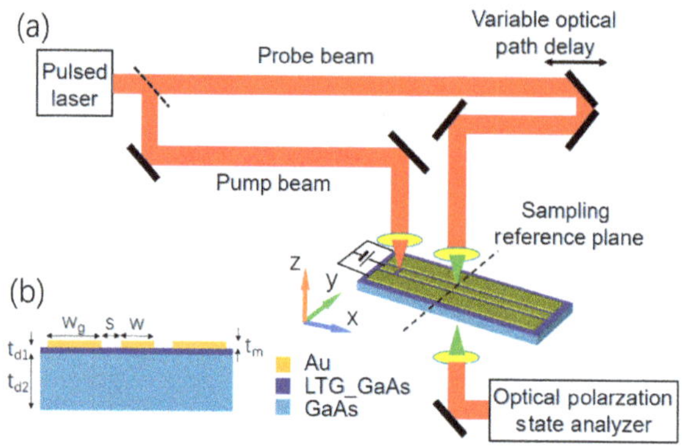

Figure 1. (**a**) The system diagram. (**b**) Cross-section of the CPW, ε_{GaAs} = 12.7, w = 45 μm, w_g = 500 μm, s = 30 μm, t_m = 0.5 μm, t_{d1} = 1 μm, t_{d2} = 200 μm. For full-wave electromagnetic analysis, the structure is enclosed in a box with the PML boundaries.

It is known that the electrical response of the gap to the carriers generated from the femtosecond laser excitation and their acceleration with a bias voltage at the electrodes is dominated by the coupled Poisson's and drift–diffusion equations for the semiconductor [18]. By solving the time and spatial-dependent electric potential and the carrier concentration, we can obtain a time-dependent photocurrent current density accounting for the femtosecond laser Gaussian envelope, which leads to the millimeter wave pulse emerging and propagating along the CPW. However, the computational cost is usually huge and needs a long time if the whole generating or detecting procedure of the voltage pulse is simulated. The complex calculation includes optical and millimeter electromagnetic wave propagation, the time-dependent carrier dynamics in the semiconductor, and their coupling procedure.

To simplify the simulating characteristics of the CPW in the on-chip system, we intend not to perform the whole calculation including the electrical response of the gap to the carriers generated from the femtosecond laser excitation. Here, we use a current source instead of directly calculating the response of the LT-GaAs to the optical femtosecond laser pulse. This laser excites a current on the gap of the center conductor represented as a current source in the simulation as shown in Figure 2a. There is also a two-port network model for the CPW in the underpart of Figure 2a. This allows us to fully describe the electrical transmissions and reflections of high-frequency devices, including a full mismatch correction. The mismatch happens when the independence of the CPW is different from the coaxial waveguide of the device under test. The current induced by the laser can be imported to commercial FEM software (version HFSS 18.1) named High-Frequency Simulator Structure (HFSS) from the former simulation results of the optical femtosecond laser pulse on the LT-GaAs [18], as illustrated in Figure 2b. With careful consideration, our investigating strategy is to use a custom monitor to obtain the average electric field on the spots of the probing laser, which represents the millimeter waveform we measured in the experiment.

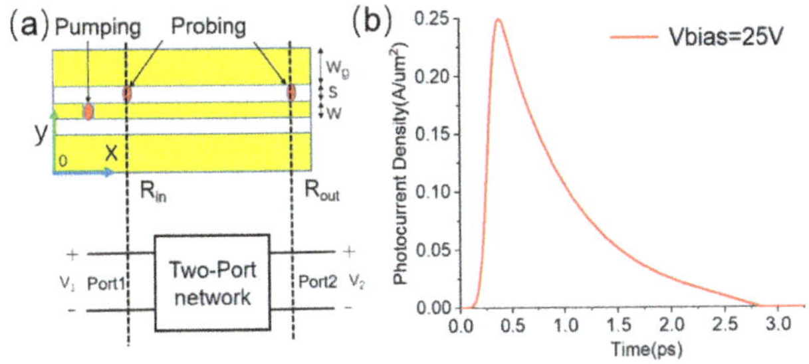

Figure 2. (**a**) The pumping and probing laser spots at different reference planes on the CPW and the two-port model based on the circuit theory. (**b**) Photocurrent density produced by Gauss shape pumping laser beam.

Theoretically, the voltage pulses generated by the laser pumping at the gap propagate along the CPW at both the $+x$ and $-x$ directions. Here, if we assume that the ports absorb all the incoming electromagnetic waves, there will be no reflections from the ports. We can obtain the waveform at different positions of the CPW by simulation. The transfer functions of the CPW can be taken by Fourier transformations of waveforms in the time domain or by ports in the frequency domain directly.

In the following part, we demonstrate the details of the forward propagating pulses at different positions on the CPW as well as the influence of the CPW geometry parameters. Finally, the transfer functions of the CPW with different lengths are calculated by the pulses at different positions.

3. Simulation Results and Discussion

Firstly, we discuss the influence of the center conductor width and slot width on the real part of the characteristic impedance Re (Z_c) and the reflection coefficient amplitude $|S_{11}|$ of the CPW ranging from 10 GHz~500 GHz. The characteristic impedance of the CPW should be 50 Ω to match the outer circuits and the probes of the vector network analyzer. Figure 3a,c show the simulation results of the real part of the characteristic impedance using the analytic model [21], and Figure 3b,d demonstrate the reflection coefficient by the FEM method for the 2-mm-long CPW without the gap. The slot width s changes from 30 μm to 60 μm with w = 45 μm for Figure 3b. The center conductor width w varies from 30 μm to 60 μm with s = 30 μm for Figure 3c,d. The Re (Z_c) increases as the w increases and approaches approximately 50 Ω when the w is about 40~45 μm within the frequency range of 10 GHz to 500 GHz. The Re (Z_c) decreases as the s increases, and it approaches approximately 50 Ω when the s is about 30~35 μm within the same frequency band. The $|S_{11}|$ approaches zero when the Re (Z_c) is close to 50 Ω. The Re (Z_c) cannot be kept at 50 Ω at each frequency at the same time. In the entire working frequency range, it should be ensured that the deviation between the Re (Z_c) and 50 Ω at the center frequency is the smallest. Here, we choose the values of w = 45 μm and s = 30 μm because in this case, the Re (Z_c) is very close to 50 Ω at 250 GHz. The ripples presented in Figure 3b,d illustrate that when the geometry of the CPW is fixed, the reflection spectra have peaks and valleys as the frequency varies. The reason for this phenomenon should be attributed to the characteristic impedance increases and decreases in a broad frequency scope. At the same time, the group delay is closely related to the characteristic impedance feature in the wide frequency domain, though the results are based on the frequency domain simulation, which can be indirectly measured by the EOS system. Our direct simulations of the waveforms are important to validate the measurement results in the experiment for these characteristic impedance features.

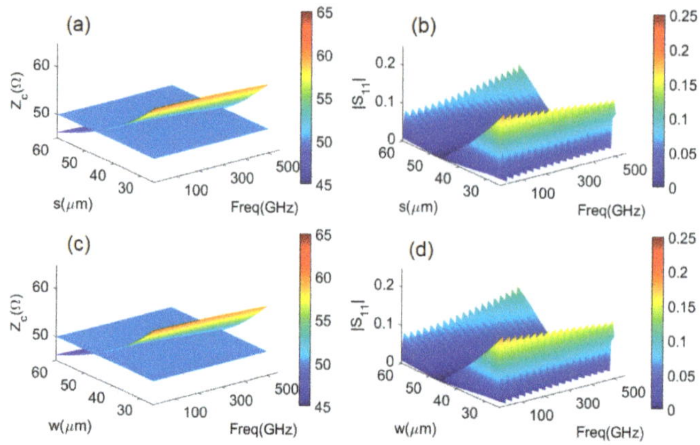

Figure 3. Simulation results of (**a**,**c**) characteristic impedance real parts and (**b**,**d**) reflection $|S_{11}|$ vary as slot width s = 30~60 μm (w = 45 μm) and center conductor width w = 30~60 μm (s = 30 μm) with the frequency ranging from 10 GHz to 500 GHz.

Secondly, we explore the voltage pulses, which can be detected by the probing laser when the geometry parameters of the CPW change. The response of the LT-GaAs on the gap to the optical femtosecond laser pulse is replaced by a current source instead of directly calculating the whole photoconductive effect. The current data are imported to HFSS, and after simulation, the integrated electric field on the spots of the probing laser is obtained. Figure 4a shows the pulse profiles at different detecting positions with x = 500 μm, 1000 μm, 1500 μm, and 2000 μm, with the gap at x = 250 μm. It is noted that the amplitudes of the pulses gradually decrease with an increase in the detecting distance, and the FWHM gradually increases from 1.4 ps to 2.6 ps. Furthermore, for the probing at x = 2000 μm, the influences of w, s, and g on the pulses has also been investigated. Figure 4b demonstrates the pulses with s = 20 μm, 25 μm, 30 μm, and 35 μm. The amplitudes of the pulses gradually decrease with an increase in the s, and the FWHM also gradually increases. Figure 4c illustrates the pulses with w = 45 μm, 50 μm, 55 μm, and 60 μm. The amplitudes of the pulses gradually decrease with an increase in the w, and the FWHM also gradually increases at the same. Figure 4d shows the pulses with the g = 5 μm, 10 μm, 15 μm, and 20 μm. The amplitudes of the pulses gradually increase with the augmentation of g, but the FWHM gradually decreases. Correspondingly, we also calculate the fast Fourier transform (FFT) of the pulses for the above-mentioned varying parameters, as depicted in Figure 5. As the width of the pulse in the time domain becomes wider, its frequency band becomes narrower. All the parameters influence the shapes of the voltage pulses, and the revealed trends are important for the EOS on-chip system design.

In order to reveal the impact of the geometry parameters of the CPW on the probed voltage pulses, a lump port is applied to the laser pumping gap, and the electrical field intensity distributions of the CPW at 100 GHz are demonstrated. Figure 6 illustrates the intensity distributions on the yox plane with z = 0 mm, the yoz plane with x = 1 mm, and the yoz plane with x = 0.25 mm (the center of the laser pumping gap). It is evident that the majority of the electrical field energy is distributed at two slots of the CPW. Furthermore, the energy is focused at the edges of slots, particularly on the sides of the center conductor, as shown in Figure 6b. Therefore, the varying of the slot width s has a greater impact on the amplitude of the millimeter wave pulses compared to the center conductor width w and the pumping gap g. The area of the electrical field intensity in the pumping gap g is smaller, resulting in a weaker effect on the pulse amplitude.

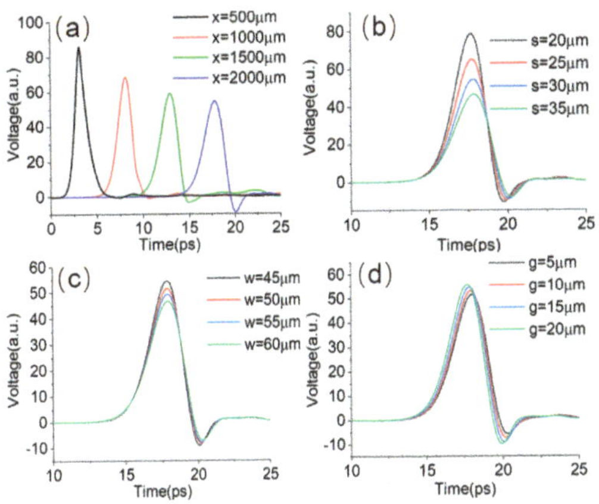

Figure 4. The pulses varying with (**a**) the position x, (**b**) the slot width s, (**c**) the center conductor width w, and (**d**) the pumping gap g with laser probing at x = 2000 μm.

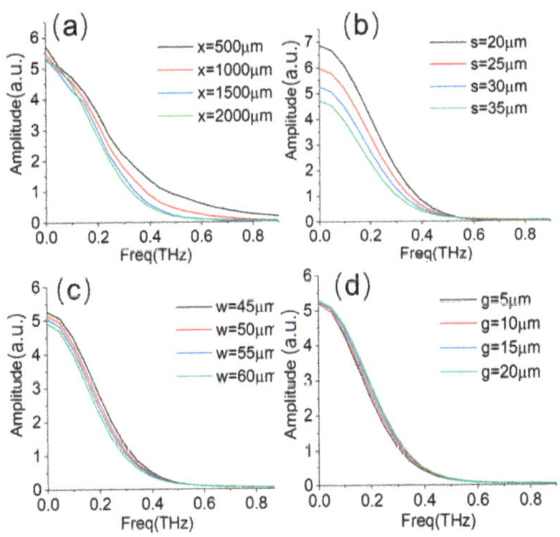

Figure 5. The FFTs of pulses varying with (**a**) the position x, (**b**) the slot width s, (**c**) the center conductor width w, and (**d**) the pumping gap g with laser probing at x = 2000 μm.

In the next step, we discuss how to obtain the transfer functions of the CPW from the pulses calculated at the different positions along the slot of the CPW, as shown in the following equation:

$$H(\omega) = \frac{V_{out}(\omega)}{V_{in}(\omega)} = \frac{FFT(v_{out}(t))}{FFT(v_{in}(t))} \quad (1)$$

where *FFT* indicates the operation of the fast Fourier transform. The time domain pulses can be measured by the EOS system in Figure 1. The transfer functions of the CPW can be acquired by Equation (1). The precise models of the CPW are very important in reducing the uncertainty of the EOS measurements for millimeter waveform metrology.

Figure 7 shows the pulse profiles and the transmission coefficients $|S_{21}|$ calculated from Equation (1) by inputting the pulse at x = 500 μm and detecting the output pulses at x = 1000 μm, 1500 μm, and 2000 μm. The corresponding lengths of the CPW are L = 500 μm, 1000 μm, and 1500 μm in the red, green, and blue colors. Those results are compared with the direction FEM calculation results of the frequency domain by a two-port model with the same lengths in Figure 6b. There is good consistency between the results from the two different methods from the time domain and frequency domain. The time domain pulses can be directly taken from the measurement results from the EOS system. This allows us to characterize the high-frequency devices with a full mismatch correction by numerically rebuilding the measured waveform with an accurate transfer function of the CPW.

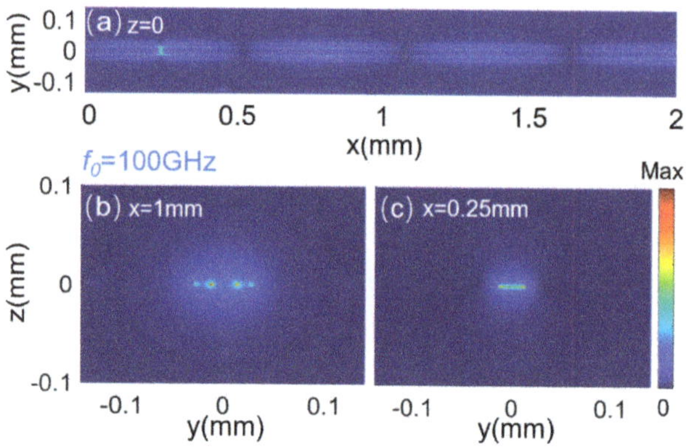

Figure 6. Electrical field intensity distributions of CPW at 100 GHz, (**a**) the yox plane with z = 0 mm, (**b**) the yoz plane with x = 1 mm, and (c) the yoz plane with x = 0.25 mm (the laser pumping gap center).

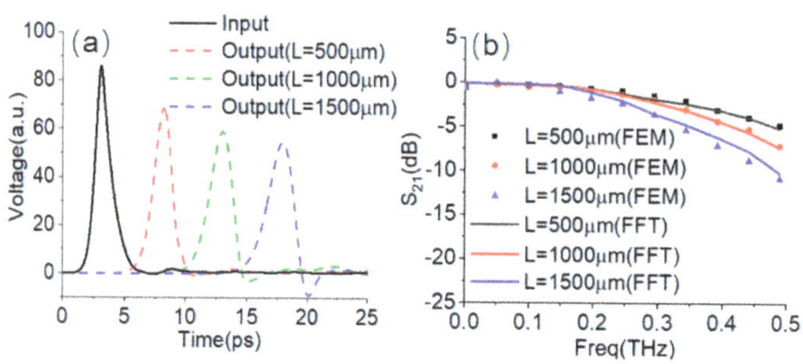

Figure 7. (**a**) The pulses as input and outputs with L = 500 μm, 1000 μm, and 1500 μm, (**b**) transmission coefficient of the CPW with pulses at different L, simulated by ports (dots) and calculated by pulses (solid lines).

4. Conclusions

In this work, based on the EOS system, an on-chip system is designed for millimeter waveform metrology with a CPW structure. The CPW is patterned on a thin layer of LT-GaAs using a GaAs substrate. The generation and detection of ultrashort pulses are realized on the CPW using one femtosecond laser. The forward propagating pulses are simulated at different positions on the CPW using a simplified current source instead of directly calculating the entire photoconductive effect. With digital signal-processing algorithms, even those temporally overlapping waveforms can be separated completely with proper known conditions. Additionally, the influence of the CPW geometry parameters on the measured pulses is discussed. The varying of the slot width has a greater influence on the amplitude of the millimeter wave pulses compared to the center conductor width and the pumping gap. This is due to the concentration of the electrical field intensity primarily occurring at the two slots of the CPW. Finally, the transfer functions of the CPW in the frequency range of 10~500 GHz with different lengths are calculated by the pulses at different positions. There is good consistency between the time domain pulse results and the direct simulations from the frequency domain with a two-port model. The above results are important for the precision measurements of the millimeter waveform and the realization of a voltage pulse standard.

Author Contributions: Conceptualization, K.Z. and X.L.; methodology, H.C. and X.L.; software, X.L.; validation, J.S. and B.L.; resources, L.L.; writing—original draft preparation, K.Z. and X.L.; writing—review and editing, D.Y.; funding acquisition, K.Z. and X.L. All authors have read and agreed to the published version of the manuscript.

Funding: This work was supported by the National Quality Infrastructure System (No. 2022YFF0605902).

Data Availability Statement: The data can be shared upon request.

Conflicts of Interest: The authors declare no conflict of interest.

References

1. Hong, W.; Jiang, Z.H.; Yu, C.; Hou, D.; Wang, H.; Guo, C.; Hu, Y.; Kuai, L.; Yu, Y.; Jiang, Z.; et al. The Role of Millimeter-Wave Technologies in 5G/6G Wireless Communications. *IEEE J. Microw.* **2021**, *1*, 101–122. [CrossRef]
2. Wang, X.; Kong, L.; Kong, F.; Qiu, F.; Xia, M.; Arnon, S.; Chen, G. Millimeter Wave Communication: A Comprehensive Survey. *IEEE Commun. Surv. Tutor.* **2018**, *20*, 1616–1653. [CrossRef]
3. van Berlo, B.; Elkelany, A.; Ozcelebi, T.; Meratnia, N. Millimeter Wave Sensing: A Review of Application Pipelines and Building Blocks. *IEEE Sens. J.* **2021**, *21*, 10332–10368. [CrossRef]
4. Wang, Z.; Chang, T.; Cui, H.-L. Review of Active Millimeter Wave Imaging Techniques for Personnel Security Screening. *IEEE Access* **2019**, *7*, 148336–148350. [CrossRef]
5. Rothbart, N.; Holz, O.; Koczulla, R.; Schmalz, K.; Hübers, H.-W. Analysis of Human Breath by Millimeter-Wave/Terahertz Spectroscopy. *Sensors* **2019**, *19*, 2719. [CrossRef] [PubMed]
6. Bieler, M.; Spitzer, M.; Pierz, K.; Siegner, U. Improved Optoelectronic Technique for the Time-Domain Characterization of Sampling Oscilloscopes. *IEEE Trans. Instrum. Meas.* **2008**, *58*, 1065–1071. [CrossRef]
7. Struszewski, P.; Pierz, K.; Bieler, M. Time-Domain Characterization of High-Speed Photodetectors. *J. Infrared Millimeter Terahertz Waves* **2017**, *38*, 1416–1431. [CrossRef]
8. Struszewski, P.; Bieler, M. Asynchronous optical sampling for laser-based vector network analysis on coplanar waveguides. *IEEE Trans. Instrum. Meas.* **2019**, *68*, 2295–2302. [CrossRef]
9. Li, E.; Yang, J.; Zhang, K.; Li, H.; Xu, Y.; Su, F.; Wang, T.; Fang, G. Systematic Study of Two-Color Air Plasma Broadband THz-TDS. *IEEE Trans. Terahertz Sci. Technol.* **2023**, *13*, 476–484. [CrossRef]
10. Williams, D.F.; Hale, P.D.; Clement, T.S.; Morgan, J.M. Calibrating electro-optic sampling systems. In Proceedings of the in 2001 IEEE MTT-S International Microwave Symposium Digest (Cat. No. 01CH37157), Phoenix, AZ, USA, 20–24 May 2001; Volume 3, pp. 1527–1530.
11. Bieler, M.; Struszewski, P.; Feldman, A.; Jargon, J.; Hale, P.; Gong, P.; Xie, W.; Yang, C.; Feng, Z.; Zhao, K.; et al. International comparison on ultrafast waveform metrology. In Proceedings of the in 2020 Conference on Precision Electromagnetic Measurements (CPEM), Denver, CO, USA, 24–28 August 2020.
12. Karpowicz, N.; Dai, J.; Lu, X.; Chen, Y.; Yamaguchi, M.; Zhao, H.; Zhang, X.-C.; Zhang, L.; Zhang, C.; Price-Gallagher, M. Coherent heterodyne time-domain spectrometry covering the entire "terahertz gap". *Appl. Phys. Lett.* **2008**, *92*, 011131. [CrossRef]
13. Feng, Z.; Zhao, K.; Yang, Z.; Miao, J.; Chen, H. 100 GHz pulse waveform measurement based on electro-optic sampling. *Laser Phys.* **2018**, *28*, 055301. [CrossRef]
14. Feng, Z.; Zhao, K.; Li, B.; Yang, Z.; Song, Z.; Cui, X. Pulse waveform characterization by electro-optic sampling at NIM. In Proceedings of the in 2020 Conference on Precision Electromagnetic Measurements (CPEM), Denver, CO, USA, 24–28 August 2020; pp. 1–2.
15. Han, C.; Wang, Y.; Li, Y.; Chen, Y.; Abbasi, N.A.; Kurner, T.F.; Molisch, A. Terahertz Wireless Channels: A Holistic Survey on Measurement, Modeling, and Analysis. *IEEE Commun. Surv. Tutor.* **2022**, *24*, 1670–1707. [CrossRef]
16. Chen, H.; Zhao, K.J.; Li, X.J.; Li, B.; Li, L. Bandwidth Improvement of Coplanar Waveguide for Electro-optic Sampling System. In Proceedings of the 2022 International Conference on Microwave and Millimeter Wave Technology (ICMMT), Harbin, China, 12–15 August 2022; pp. 1–3.
17. Bieler, M.; Fuser, H.; Pierz, K. Time-Domain Optoelectronic Vector Network Analysis on Coplanar Waveguides. *IEEE Trans. Microw. Theory Tech.* **2015**, *63*, 3775–3784. [CrossRef]
18. Burford, N.; El-Shenawee, M. Computational modeling of plasmonic thin-film terahertz photoconductive antennas. *J. Opt. Soc. Am. B* **2016**, *33*, 748–759. [CrossRef]
19. Nguyen, T.K.; Kim, W.T.; Kang, B.J.; Bark, H.S.; Kim, K.; Lee, J.; Park, I.; Jeon, T.-I.; Rotermund, F. Photoconductive dipole antennas for efficient terahertz receiver. *Opt. Commun.* **2016**, *383*, 50–56. [CrossRef]

20. Bieler, M.; Spitzer, M.; Hein, G.; Siegner, U. Time-domain characterisation of non-coplanar high-frequency components up to 300 GHz. *Electron. Lett.* **2002**, *38*, 1038–1039. [CrossRef]
21. Heinrich, W. Quasi-TEM description of MMIC coplanar lines including conductor-loss effects. *IEEE Trans. Microw. Theory Tech.* **1993**, *41*, 45–52. [CrossRef]

Disclaimer/Publisher's Note: The statements, opinions and data contained in all publications are solely those of the individual author(s) and contributor(s) and not of MDPI and/or the editor(s). MDPI and/or the editor(s) disclaim responsibility for any injury to people or property resulting from any ideas, methods, instructions or products referred to in the content.

electronics

MDPI

Article

Multi-Link Prediction for mmWave Wireless Communication Systems Using Liquid Time-Constant Networks, Long Short-Term Memory, and Interpretation Using Symbolic Regression

Vishnu S. Pendyala [1,*,†] and Milind Patil [2,†]

1 Department of Applied Data Science, San Jose State University, San Jose, CA 95192, USA
2 OSI Engineering Inc., 901 Campisi Way, Suite-160, Campbell, CA 95008, USA
* Correspondence: vishnu.pendyala@sjsu.edu
† These authors contributed equally to this work.

Abstract: A significant challenge encountered in mmWave and sub-terahertz systems used in 5G and the upcoming 6G networks is the rapid fluctuation in signal quality across various beam directions. Extremely high-frequency waves are highly vulnerable to obstruction, making even slight adjustments in device orientation or the presence of blockers capable of causing substantial fluctuations in link quality along a designated path. This issue poses a major obstacle because numerous applications with low-latency requirements necessitate the precise forecasting of network quality from many directions and cells. The method proposed in this research demonstrates an avant-garde approach for assessing the quality of multi-directional connections in mmWave systems by utilizing the Liquid Time-Constant network (LTC) instead of the conventionally used Long Short-Term Memory (LSTM) technique. The method's validity was tested through an optimistic simulation involving monitoring multi-cell connections at 28 GHz in a scenario where humans and various obstructions were moving arbitrarily. The results with LTC are significantly better than those obtained by conventional approaches such as LSTM. The latter resulted in a test Root Mean Squared Error (RMSE) of 3.44 dB, while the former, 0.25 dB, demonstrating a 13-fold improvement. For better interpretability and to illustrate the complexity of prediction, an approximate mathematical expression is also fitted to the simulated signal data using Symbolic Regression.

Keywords: liquid neural networks; extremely high frequency; mmWave; 5G network; genetic programming

1. Introduction

Wireless communication systems, particularly the 5G and the upcoming 6G networks, are increasingly adopting the extremely high-frequency (EHF) millimeter-wave (mmWave) and tremendously high-frequency (THF) sub-terahertz T-waves. The wavelength of the mmWave frequencies is between one centimeter and one millimeter, hence the name. They offer a large amount of available spectrum, greater capacity, and more bandwidth than traditional bands. This means that they can be used to provide very high data rates. The 5G and 6G wave technology for mobile network operators enable many important applications such as autonomous driving [1], precision agriculture [2], mobile virtual reality [3], and high-definition video broadcasting [4]. They offer faster deployment and higher return on investment (ROI). Multi-link prediction refers to the process of forecasting the quality of potential connections between a user device and a base station. Unlike traditional cellular networks where there is a single connection between a device and a tower, mmWaves use narrow, steerable beams and follow a process called beamforming. Therefore, connections along multiple paths can be formed simultaneously. The quality of the connection is often measured by the Signal-to-Noise Ratio (SNR). Higher values indicate a stronger signal relative to background noise. By predicting the quality, the device can seamlessly switch to the predicted high-quality link without losing the connection.

However, owing to their short wavelength, EHF and THF waves have a strict Line of Sight requirement. This means that they cannot travel through walls or trees and are highly sensitive to obstacles. The channel changes very quickly as the user moves. Wireless systems using them are impacted by fluctuating channel quality when the beam propagates in various directions simultaneously. Millimeter-wave communication is also not energy efficient. Slight deviations in the path of the transceivers or the presence of interfering elements can cause significant alterations in link quality unpredictably. Low-latency applications necessitate the ability to accurately forecast the link quality across different cells and directions. Previous attempts to address the challenge of predicting multiple links using AI/ML techniques involved certain Recurrent Neural Network (RNN) models like Long short-term memory (LSTM) for the link prediction [5].

Link prediction is the important task of estimating the quality of the signal between nodes in a network. It plays an important role in EHF and THF systems because these signals are inherently flaky and unpredictable, but the low-latency applications require accurate prediction. Link prediction helps in optimizing network performance, making routing decisions, and resource allocation.

Traditionally, the parameters for Artificial Neural Networks (ANNs), including RNNs like LSTM are fixed after training. ANNs use these fixed parameters to make decisions, a process called inferencing. Liquid Neural Networks (LNNs) are a better type of RNN that can learn even at the time of decision-making and are therefore better suited to real-world applications [6]. LNNs are a progression of Neural Ordinary Differential Equations (ODEs). Neural ODEs [7] use an approximation to evolve the hidden state by using a fixed time interval or a "time constant". Liquid Time-Constant Networks or LTCs are a special type of LNN that use a time constant that is not fixed, giving it the flexibility to dynamically adapt to changes in the data [8], which for this work is the wireless EHF and THF signal. We hypothesize that LTCs are better suited to the problem of multi-link prediction for mmWave and sub-terahertz because of the highly unpredictable nature of the signal, and the experiments detailed later confirm this. LTCs adapt to the rapid fluctuations in the data even during inferencing [9]. Their behavior is more explainable, they can learn from smaller amounts of data, and are computationally effective, requiring a smaller number of neurons [8]; hence, they can run on edge devices like smartphones.

The simulation in this work is based on the Urban Microcellular Infrastructure (UMI) Channel model [10], an indoor scenario that typically uses 28 GHz as the operating carrier frequency. Therefore, the EHF and THF waves are generated for the experimental setup using a real-time simulation involving multiple links with trajectories of moving individuals and obstructing vehicles in a scenario set at 28 GHz. The simulation uses appropriate noise figure and path loss encompassing both Line-of-Sight (LOS) and Non-Line-of-Sight (NLoS) propagation paths following the 3GPP Channel model [11]. The parameters set for the simulation are as specified in Table 1. Also, the current literature [5] uses 28 GHz for experiments with LSTM. Using the same 28 GHz frequency for this work helps ensure a fair evaluation of the results.

The experiments hypothesize that more precise and reliable predictions regarding the link quality in complex wireless communication scenarios can be achieved using LTCs. The proposed LTC-based Multi-Link Prediction model seeks to enhance the accuracy and efficiency of link quality predictions by leveraging recent developments in artificial intelligence technologies. The experiments are intended to validate the effectiveness of LTCs in a practical simulation environment to confirm the hypothesis that LTCs address the challenges associated with predicting link quality in dynamic wireless communication systems. This research contributes to the evolution of predictive modeling in wireless communication, paving the way for robust and adaptive systems in the future as wireless communication systems evolve towards 6G networks and beyond.

Table 1. NYUSIM simulation parameters for generating the data.

Parameter	Value
Environment	Urban microcell indoor
Area	200 × 200 m^2
Carrier frequency	28 GHz
Path loss model	5G urban microcell indoor
Number of base station antennas	64
Number of user equipments	8
Bandwidth	400 MHz
Radius of base station	100 m
Height of base station	10 m
Height of user equipment	1.7 m
Blocker dimensions	1.7 m × 0.3 m
Transmission power	23 dBm
Noise figure	9 dB
Sampling interval	20 ms

In the case of mmWave frequencies, it is expected that the distance between cellular base stations remains consistent. With carrier frequencies progressing towards terahertz, the wider bandwidth of the channel allows for covering similar distances. This is achieved through the exponential growth in antenna gains as frequencies rise, assuming the physical antenna remains unchanged [12].

The LSTM model is impacted by certain shortcomings when used for link prediction for EHF and THF. LSTM outcomes tend to be less accurate, particularly in scenarios where prediction involves data that are not continuous. EHF and THF waves are prone to attenuation due to the oxygen and other matter in the atmosphere. The waves are absorbed by certain compounds such as water vapor in the atmosphere. The higher the frequency, the greater the absorption. Also, as mentioned earlier, EHF and THF waves propagate only along the Line-of-Sight paths. They cannot bend around obstacles. Due to the varying distances between the transmitter, receiver, and any obstructing objects present on each link, the signal could be choppy. Consequently, for such scenarios, when the LSTM model is applied, the results may not be accurate to the desired extent. In scenarios where there are more links between the transmitter and the receiver, the effectiveness of an LSTM model might diminish, resulting in suboptimal predictions [13,14].

The overall mobile handover operation (HO) probability with respect to both horizontal as well as vertical hand-off needs to be characterized to attain specific mobile coverage probability in mmWave and sub-terahertz networks [15].

1.1. Related Work

A combination of Convolutional Neural Network (CNN) and Recurrent Neural Network (RNN) deep learning models have been used to predict future blockages and beams for mmWave systems. This helps with proactive handovers between base stations to ensure uninterrupted connectivity for users [16]. Diverse approaches have been tried for link quality prediction, including using camera images [17]. Predicting link blockages is important to ensure seamless connectivity. The predictions help to proactively switch beams and handoff (HO). Computer vision has been used to conduct these predictions based on camera images in yet another work [18]. Transfer learning using deep neural networks was attempted to predict the optimal beams for multi-links, resulting in reduced interference and training overhead [19]. EHF waves suffer from low spectral efficiency, narrow coverage, and difficulty in Non-Line-of-Sight (NLoS) propagation. It is therefore important to model path loss accurately for optimal base station placement. There are three types of path loss modeling methods: empirical, deterministic, and machine learning-based. Machine learning-based modeling uses measured data to train a model to predict path loss. A novel machine learning scheme called multi-way local attentive learning has been proposed to model and predict path loss [20].

By understanding the environment using a combination of analytical models such as geometric analysis to recognize the shadowed regions that separate Line-of-Sight (LoS) and Non-Line-of-Sight (NLoS) scenarios and deep neural networks, researchers have built systems that can proactively allocate resources for better performance [21]. The allocation is based on link quality predictions as described in this work but using deep neural networks for regression. Not just deep learning, but traditional machine learning techniques such as Support Vector Regression (SVR), Random Forest (RF), and Gradient Tree Boosting (GTB) have also been used for the task of mmWave link prediction [22]. In yet another novel approach, researchers extract a sparse feature representation using non-deterministic quantization and apply deep neural network (DNN) to learn from those features for mmWave beam prediction [23]. Channel State Information (CSI) is fused with information about the user's location to predict the beam [24]. The prediction is still carried out using a neural network that adopts Adjustable Feature Fusion Learning (AFFL).

Another approach [25] combines Deep Neural Networks (DNNs) with Long Short-Term Memory (LSTM) networks to create a new prediction method. This approach considers both past channel information and the position of the device. The proposed method can predict both large-scale trends and small-scale fluctuations in mmWave channel features. The approach achieved over 4.5% improvement in accuracy compared to existing approaches. To detect the motion of blockers such as a walking person close to the Line-of-Sight (LoS) path, an mmAlert system was proposed [26] using the passive sensing technique. It could predict 90% of the LoS blockage, with a sensing time of 1.4 s being sufficient enough to provide a timely warning. Privacy is important in such applications. To preserve privacy, instead of cameras, point clouds have been used for predicting signal strength in millimeter-wave communication systems [27]. Point clouds are 3D representations of spaces. While cameras may capture sensitive information, point clouds are devoid of privacy concerns. The approach still achieves accuracy that is comparable to traditional image-based methods.

As can be seen, link prediction has been attempted via several methods, mostly using deep learning through Artificial Neural Networks (ANN) and some traditional machine learning approaches. The accuracy of the prediction is reasonable but, as this work confirms, LTC predictions are far more accurate than with LSTM, which so far was the most relevant deep learning framework for the problem.

1.2. Contribution

To our knowledge, this work is unique in achieving a significantly better prediction of the Signal-to-Noise Ratio (SNR) of EHF signals using Liquid Time-Constant Networks. Explainability and interpretability play an important role in machine learning [28]. This work is also novel in applying Symbolic Regression to fit a mathematical expression to SNR values of a simulated EHF wireless system for interoperability. The core research contributions can be summarized as follows:

- Quantitative demonstration of substantial improvement in multi-link prediction for mmWave wireless communication systems using Liquid Time-Constant Networks (LTC) over conventional methods such as using Long Short-Term Memory.
- Interpretation of the SNR values of mmWave signal using Symbolic Regression.

The rest of the paper is organized as follows. Section 2 details the approach followed for the experiments. It briefly explains the concepts, the dataset used, how it was generated, and the setup for Symbolic Regression. Numerical experiments, including the hyperparameter settings, and results are discussed in Section 3. Section 4 discusses the results from the experiments and the conclusions are included in Section 5.

2. Materials and Methods

For simulating the mmWave and sub-terahertz wireless channels, an open-source simulator developed by the New York University called NYUSIM is used. The product can simulate many real-life environments such as urban macrocell, urban microcell, indoor hotspot, and rural macrocell. The performance of a simulated communication channel is measured by the Signal-to-Noise Ratio (SNR). The SNR is a measure of the quality of the received signal and is defined as the ratio of the power of the received signal to the power of the noise present in the channel [29]. The data generated using the simulator are then used for the experiments with LSTM and LTC. The results from both are compared to test the hypothesis that LTC achieves better prediction. A mathematical expression is also fitted to the data using Symbolic Regression through genetic programming [30]. The mathematical expression gives insights into the nature of variations in the SNR values.

2.1. Long Short-Term Memory—LSTM

For comparison, the results from using LSTM [31] for link quality prediction are used as the baseline. The training part is performed using a dataset that has Signal-to-Noise (SNR) values generated during simulation trials, enabling the model to decrypt the underlying patterns and links. The input layer of the LSTM network takes the SNR value sequence, followed by a hidden layer, and then an output layer. The LSTM cells in the model function in a unidirectional manner, aiding in the sequential processing of data. The training of the model utilizes back-propagation through time, which efficiently propagates error gradients through the LSTM cells over temporal spans. The results highlight the effectiveness and assurance of LSTM models in tasks associated with the prediction of new SNR values as the signal propagates.

2.2. LTC

The experiments are then repeated using LTC instead of LSTM. The SNR values are now passed as inputs to LTC. LTCs are also a type of Recurrent Neural Networks (RNN). Unlike the traditional RNNs, which process information in discrete steps, LTC networks handle information that is more unpredictable over time. Mathematically, they operate using differential equations, allowing them to model systems that evolve more dynamically [8]. A key feature of LTC is the concept of a "liquid time constant", from which LTC gets its name. This refers to the fact that each neuron in the network has its internal timescale for processing information. This flexibility allows the network to capture more granular rates of change in the data even at the time of inferencing.

2.3. Dataset: Simulation

The dataset comprises Signal-to-Noise Ratio (SNR) measurements calculated every 20 ms. SNR is computed by comparing the received signal power to the noise power in the channel. The efficiency of a telecommunications system model operating in an urban microcell indoor environment was examined in this work. A frequency of 28 GHz was used to generate the SNR values in the dataset. The parameters for the environment setup used for the simulation are as shown in Table 1. The parameters are self-explanatory.

The large-scale path loss model used in NYUSIM is a Close-In free space reference distance (CI) model with a 1 m reference distance [32]. It includes an extra attenuation term due to various atmospheric conditions. The model is given by Equation (1).

$$PL^{CI}(f,d)[dB] = FSPL(f,1m)[dB] + 10nlog_{10}(d) + AT[dB] + \chi\sigma, \quad (1)$$

where $PL^{CI}(f,d)[dB]$ is the path loss in dB at a distance of d meters and carrier frequency of f GHz;
$FSPL(f,1m)[dB]$ is the free space path loss in dB at a distance of 1 m and carrier frequency of f GHz;
n is the path loss exponent (n = 2 for free space);
$AT[dB]$ is the total atmospheric absorption term;
$\chi\sigma$ is the shadow fading (SF) that refers to the signal attenuation due to obstacles in the Line of Sight, modeled as a log-normal random variable with zero mean;
σ is the standard deviation in dB.

The Free Space Path Loss (FSPL), an idealized theoretical concept that refers to the attenuation of radio signal strength as it travels through space, with no obstacles or reflections, is given by Equation (2).

$$FSPL(f, 1m)[dB] = 20log_{10}(\frac{4\pi f \times 10^9}{c}) = 32.4[dB] + 20log_{10}(f), \tag{2}$$

where c is the speed of light.

The path loss exponent is a parameter that describes how the path loss increases with distance. For free space, the path loss exponent is 2. However, the path loss exponent can vary depending on the environment. For example, in urban areas, the path loss exponent is typically greater than 2 due to the presence of buildings and other obstructions. The total atmospheric absorption term is a measure of the attenuation of the signal due to absorption by the atmosphere. The absorption is caused by the interaction of the radio waves with the molecules in the atmosphere. The absorption is frequency-dependent, with higher frequencies being absorbed more than lower frequencies [33].

Shadow fading is a random variation in the received signal strength due to changes in the propagation environment. Shadow fading is caused by changes in the terrain, buildings, and other objects between the transmitter and receiver. Shadow fading is typically modeled as a log-normal random variable [29] as can be seen from Equation (1).

The large-scale path loss model is used to predict the received signal strength at a given distance from the transmitter. The model can be used to design wireless networks and to estimate the coverage area of a wireless network [33].

2.4. Dataset Generation

The program to generate the dataset of SNR values using the simulator is written in MATLAB R2022b Update 9 (9.13.0.2553342), 64-bit (maci64). The SNR values are produced by simulating an mmWave wireless communication system operating at a carrier frequency of 28 GHz, encompassing both Line-of-Sight (LOS) and Non-Line-of-Sight (NLoS) propagation paths. The procedure emulates a Channel Impulse Response (CIR) for each trajectory, followed by multiplication with a random input signal to produce a received signal. The SNR is calculated as the ratio of signal power to noise power, assuming noise power is directly proportional to signal power, and a fixed SNR of 30 dB is maintained.

2.5. SNR Calculation

There are several imperative tools used for classifying the various patterns of mobility in wireless communications. These patterns are known as mobility models and include the Vehicular Mobility Model, the High-Speed Train Mobility Model, the Human Mobility Model, and the Ship Mobility Model. Through the extraction of extensive datasets, such mobility models provide researchers with the ability to scrutinize and approximate the influence of several mobility factors such as vehicular speed, congestion levels, ambiguity, social interactions, location preferences, and more. To perform an extensive experiment in this field, it is critically important to assess both the Human Mobility Model (HMM) and the Vehicular Mobility Model (VMM) as they play a vital role in apprehending and speculating the dynamics of wireless communications. By concentrating on such specific mobility models, researchers can attain valuable insights into the complexities of mobility patterns and enhance the overall efficiency and reliability of wireless communication systems [5].

Using a simulated channel model, an assessment has been conducted to determine the received signal power and noise power at the user equipment. The received signal power is calculated as the product of the transmitted power, the antenna gain, and the path loss between the base station and the User Equipment (UE). The noise power is calculated as the product of the noise figure, the bandwidth, and the Boltzmann constant. Then, the Signal-to-Noise Ratio (SNR) is calculated as the ratio of the power of the received signal to the power of the noise, expressed in decibels (dB) [34].

The Signal-to-Noise Ratio (SNR) is calculated as the ratio of the power of the received signal to the power of the noise, expressed in decibels (dB). In the context of this problem, the SNR can be calculated as

$$SNR = 10log_{10}(P_{signal}/P_{noise})$$

where P_{signal} is the power of the received signal and P_{noise} is the power of the noise.

The parameters required to calculate the SNR for this problem are as follows. P_{signal}: The power of the received signal, which is calculated as the product of the channel gain and the transmitted power. P_{noise}: The power of the noise, which is calculated as the product of the noise figure and the receiver bandwidth [5].

The distance between the transmitter and receiver, the carrier frequency, and the properties of the propagation environment determine the channel gain. The transmitter power and receiver bandwidth are also specified in the problem. The noise figure represents the noise added by the receiver and is also specified in the problem [5].

Subsequently, an analysis has been conducted to establish the SNR by comparing the power of the received signal to the power of the noise, resulting in creating a dataset comprising SNR values recorded at 20 millisecond intervals. These recorded SNR values offer significant insights into the operational efficiency of the communication system within an indoor microcellular setting in urban areas [33].

2.6. Genetic Programming Based Symbolic Regression

The dataset is modeled using a mathematical expression to gain deeper insights into the SNR values generated and predict future SNR values using the mathematical expression. To find a mathematical expression that best fits a given dataset, Symbolic Regression is applied to the dataset. It is a type of regression analysis in which the goal is to find a mathematical expression that best fits a given set of data points. In contrast to traditional regression methods that use predefined functions such as linear or polynomial equations, Symbolic Regression allows the model to discover its own functional form by searching through a space of mathematical expressions. Genetic programming is a type of evolutionary algorithm that is used to find solutions to complex problems by mimicking the process of natural selection. In genetic programming, a population of candidate solutions, known as "individuals" is evolved over multiple generations using principles of genetic variation and selection. The fitness function is a measure of how well a particular individual—a potential solution—fits the data.

In Symbolic Regression, the fitness function is typically based on the degree of error between the actual data points and the predicted values generated by the individual's mathematical expression. Crossover and mutation are two key genetic operators used in genetic programming. Crossover involves combining genetic material from two parent individuals to create a new offspring individual, while mutation involves randomly altering the genetic material of an individual in order to introduce new variations. Tournament selection is a common method of selecting individuals for reproduction in genetic programming. In tournament selection, a subset of individuals from the population is chosen at random, and the individual with the highest fitness within that subset is selected for reproduction. In genetic programming, individuals are represented as trees of nodes. Terminal nodes represent input variables or constants, while non-terminal nodes represent mathematical operations or functions [35].

The above experiment has been performed on the dataset that was generated by an earlier-simulated model consisting of SNR values. There are predefined arithmetic functions such as add, sub, mul sin, cos, etc., for the Symbolic Regression. The code is implemented in Python. The SymbolicRegressor class is imported from gplearn.genetic package. Custom-defined functions such as ReLU were defined and added to the function set to provide the algorithm with more flexibility in modeling the data. This can be achieved using the make_function() method. The other parameters that the SymbolicRegressor constructor can take to configure the model are as follows:

1. population_size: The number of individuals (mathematical expressions) in each generation.
2. function_set: The set of functions and terminals that can be used in the mathematical expressions.
3. generations: The maximum number of generations to evolve the population.
4. stopping_criteria: The threshold value for the Mean Absolute Error (MAE) to stop evolution if it falls below this value.
5. p_crossover, p_subtree_mutation, p_hoist_mutation, p_point_mutation: The probabilities of applying crossover, subtree mutation, hoist mutation, and point mutation operations during evolution.
6. max_samples: The maximum proportion of samples to use in each generation during fitness evaluation.
7. verbose: Whether to enable verbose output during evolution.
8. parsimony_coefficient: A coefficient to balance between the goodness of fit and the complexity (parsimony) of the mathematical expressions.
9. random_state: The random seed for reproducibility.

These parameters allow us to customize the SymbolicRegressor model to fit the specific needs. For example, one can increase the population_size to improve the accuracy of the model, or these values could also be reduced generations to speed up the training process. After passing the dataset through the Symbolic Regressor, the symbolic expression obtained for the dataset is as follows.

$$
\begin{aligned}
\text{next\_e} = {} & \sin(0.0667X_0) \\
& - \sin\Bigg(\frac{30.332}{X_0} + \frac{1}{X_0^2} \cdot \{135.039 \cdot \max(0, -0.077X_0 \\
& + \sin(0.077X_0 + \sin(0.077X_0 + 0.2009)) \\
& + \frac{14.385}{X_0})\}\Bigg) \\
& + \sin(0.0667X_0 + 1) \\
& + \sin(0.077X_0 - 0.563 + \frac{14.385}{X_0}) \\
& + \sin(0.077X_0 + \sin(0.077X_0 - 0.579) - 0.338) \\
& - \sin(\sin(0.0667X_0 + \sin(0.077X_0 - 0.579) \\
& \quad - \max(0, -0.077X_0) - 15.408 + \frac{10.418}{X_0}\Big)\Big) \\
& - 14.595
\end{aligned}
$$

Also, the above expression can be depicted as a tree diagram using the graphviz python package as shown in Figure 1. Graphviz helps visualize the complex math expressions discovered by Symbolic Regression as tree diagrams, making them easier to understand and analyze. The leaves of the math parse tree [36] are numeric constants and variables from the symbolic expression. The internal nodes of the tree are arithmetic operations such as addition, subtraction, multiplication, division, and sine. Traversing the tree to its root will generate the entire symbolic expression. The tree diagram is a visual indication of the complexity of the regression.

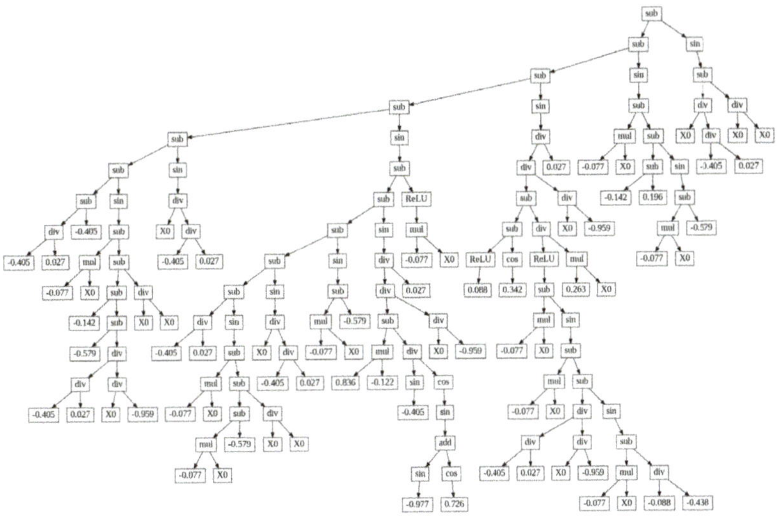

Figure 1. The mathematical expression from the Symbolic Regression. Modeling the SNR values is depicted as a tree diagram for better visualization.

Now, when the trained model is evaluated for Symbolic Regression, the fit of the model to the data is determined by R2 score. R2 score is also called as the coefficient of determination. It is a statistical measure that tells how well a regression model, in this case the mathematical expression, fits the available data. It quantifies the proportion of variance in the dependent variable, the SNR values in this case, that is explained by the model. It ranges from 0 to 1, with 0 being the worst fit and 1 being the perfect fit. A higher R2 score means that the model is better at predicting the dependent variable. The evaluation of the model is illustrated in a scatter plot shown in Figure 2. The x-axis is the timeline indexed by the observation count. The y-axis refers to the Signal-to-Noise Ratio (SNR). The blue dotted line represents the observed SNR values. The red dotted line represents the symbolic function, which is the mathematical formula discovered by the Symbolic Regression algorithm to approximate the true function. The closer the red line is to the blue line, the better the fit.

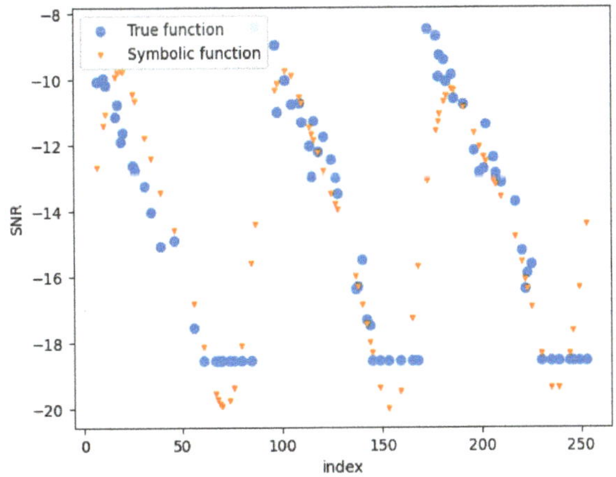

Figure 2. The scatter plot from the Symbolic Regression shows how well the mathematical expression fits the data.

As can be seen, the fit is reasonably satisfactory. The R2 score for this fit is 0.82497861499. The symbolic function captures the general trend of the true function reasonably well. However, there are some deviations between the two lines, particularly at higher and lower SNR values. The deviation increases when the SNR is below −18 dB and above −10 dB. The deviation at lower SNR values may be because of the noise floor, which sets a minimum limit on how low the SNR can go [37]. In mmWave systems, this noise floor can be caused by thermal noise in the components, atmospheric noise, and interference from other sources. The short plateaus of the signal at the bottom may be caused by the noise floor. On the other hand, the deviation at higher SNR values can be related to non-linear effects in amplifiers and other components that can distort the signal, reducing the effective SNR even if the raw signal power is high.

The fit method of the SymbolicRegressor class uses genetic programming to evolve a population of mathematical expressions that best fit the training data. The predict method uses the trained model to predict the output values for the test features. The plot shown in the figure below the true target values (y_test) against the test features (X_test) as points labeled "True function", and the predicted values (y_gp1) against the test features as points labeled "Symbolic function". Here is a more detailed explanation of each step:

1. fit method: The fit method uses genetic programming to evolve a population of mathematical expressions. Genetic programming is a type of evolutionary algorithm that uses natural selection to find the best fit for a given set of data. In this case, the data are the training set, and the goal is to find a mathematical expression that can predict the target variable for any given set of features.
2. predict method: The predict method uses the trained model to predict the output values for the test features. The test features are a set of data that were not used to train the model. The model uses the mathematical expression that it evolved during the fit method to predict the target variable for each test feature.
3. Plot: The plot shows the true target values (y_test) against the test features (X_test) as points labeled "True function". The plot also shows the predicted values (y_gp1) against the test features as points labeled "Symbolic function". The two sets of points should be close together, which indicates that the model was able to learn the relationship between the features and the target variable.

3. Link Prediction Experiments and Results

The LSTM experiment hyper-parameters are shown in Table 2. These are chosen based on intuition from working with similar problems in the past.

Table 2. LSTM Hyper-parameters.

Parameter	Value
Input size	1
Hidden size	32
Number of layers	2
Output size	1
Normalization	MinMaxScaler
Optimizer	Adam
Learning rate	0.001
Number of epochs	1000

Results for the LSTM experiment are presented in Table 3.

Table 3. LSTM Results at 28 GHz.

Evaluation Metric	Value
Validation RMSE	3.9762
Test RMSE	3.4490

Improvements with Liquid Time-Constant Network

The best results with LTC were obtained in epoch 197. The training loss obtained in this epoch is 0.04, training MAE (Mean Absolute Error) is 0.16, validation loss is 0.16, validation MAE (Mean Absolute Error) is 0.38, validation RMSE (Root Mean Square Error) is 0.41, testing loss is 0.06, testing MAE (Mean Absolute Error) is 0.20, and test RMSE (Root Mean Square Error) is 0.25.

The RMSE values for the LTC experiment are given in Table 4.

Table 4. LTC Results at 28 GHz.

Evaluation Metric	Value
Validation RMSE	0.41
Test RMSE	0.25

4. Discussion

As can be seen from the RMSE values from the experiments with the LSTM and LTC frameworks, LTC performs thirteen times better than the LSTM framework when it comes to link prediction. The RMSE obtained using LSTM in this work, 3.44 dB, is comparable with that in the literature, which is 3.14 and 2.84 for two different use cases [5]. The results can be interpreted from the Symbolic Regression perspective. As can be seen from the complexity of Figure 1 and the corresponding mathematical expression, the SNR values fluctuate substantially, as is characteristic of mmWaves. LSTM is not good at modeling this level of randomness, at least in comparison with LTC, confirming the initial hypothesis.

It must also be noted that Symbolic Regression has been proven to be NP-hard [38]. There is no known efficient, polynomial-time algorithm to solve it for all cases. As the problem size increases, the computational cost grows exponentially, making it increasingly difficult to solve in a reasonable amount of time. Even with the approximate Symbolic Regression implementation in this work, the computational complexity is high due to the vast search space, evaluation requirements, and limitations of heuristic approaches. On the other hand, LSTM is local in space and time with a linear computational complexity per time step [31] and LTC is similar.

5. Conclusions

The mmWaves that characterize 5G networks pose some unique challenges in modeling, even using the most effective machine learning algorithms that have been employed so far in the literature. This work attempted to better model the SNR values from a simulated mmWave propagation using LTC and achieved outstanding results that are thirteen times better than the results from the best framework used in the literature, which is LSTM. Using LSTM resulted in a test Root Mean Squared Error (RMSE) of 3.44 dB, and LTC, 0.25 dB. The intricacy of the prediction task that makes LTC more suitable is confirmed by the interpretation using Symbolic Regression, which resulted in a complex arithmetic expression to model the fluctuations in the SNR values. A future direction is to use other recent and interpretable deep learning frameworks, such as Kolmogorov–Arnold Networks.

Author Contributions: Conceptualization, V.S.P.; methodology, V.S.P.; software, M.P.; validation, M.P.; formal analysis, V.S.P.; investigation, V.S.P. and M.P.; resources, M.P. and V.S.P.; data curation, M.P.; writing—original draft preparation, M.P. and V.S.P.; writing—review and editing, V.S.P.; visualization, M.P.; supervision, V.S.P.; project administration, V.S.P. All authors have read and agreed to the published version of the manuscript.

Funding: This research received no external funding

Institutional Review Board Statement: Not applicable.

Informed Consent Statement: Not applicable.

Data Availability Statement: The raw data supporting the conclusions of this article will be made available by the authors on request.

Conflicts of Interest: Author Milind Patil was employed by the company OSI Engineering Inc. The remaining authors declare that the research was conducted in the absence of any commercial or financial relationships that could be construed as a potential conflict of interest.

References

1. Fan, L.; Wang, J.; Chang, Y.; Li, Y.; Wang, Y.; Cao, D. 4D mmWave radar for autonomous driving perception: A comprehensive survey. *IEEE Trans. Intell. Veh.* **2024**, *9*, 4606–4620. [CrossRef]
2. Usman, M.; Ansari, S.; Taha, A.; Zahid, A.; Abbasi, Q.H.; Imran, M.A. Terahertz-based joint communication and sensing for precision agriculture: A 6g use-case. *Front. Commun. Netw.* **2022**, *3*, 836506. [CrossRef]
3. Le, T.T.; Van Nguyen, D.; Ryu, E.S. Computing offloading over mmWave for mobile VR: Make 360 video streaming alive. *IEEE Access* **2018**, *6*, 66576–66589. [CrossRef]
4. Nallappan, K.; Guerboukha, H.; Nerguizian, C.; Skorobogatiy, M. Live streaming of uncompressed HD and 4K videos using terahertz wireless links. *IEEE Access* **2018**, *6*, 58030–58042. [CrossRef]
5. Shah, S.H.A.; Sharma, M.; Rangan, S. LSTM-Based Multi-Link Prediction for mmWave and Sub-THz Wireless Systems. In Proceedings of the ICC 2020—2020 IEEE International Conference on Communications (ICC), Dublin, Ireland, 7–11 June 2020; pp. 1–6. [CrossRef]
6. Chahine, M.; Hasani, R.; Kao, P.; Ray, A.; Shubert, R.; Lechner, M.; Amini, A.; Rus, D. Robust flight navigation out of distribution with liquid neural networks. *Sci. Robot.* **2023**, *8*, eadc8892. [CrossRef]
7. Chen, R.T.; Rubanova, Y.; Bettencourt, J.; Duvenaud, D.K. Neural ordinary differential equations. *Adv. Neural Inf. Process. Syst.* **2018**, *31*, 6571–6583.
8. Hasani, R.; Lechner, M.; Amini, A.; Rus, D.; Grosu, R. Liquid time-constant networks. In Proceedings of the AAAI Conference on Artificial Intelligence, Virtual, 2–9 February 2021; Volume 35, pp. 7657–7666.
9. Gajjar, P.; Saxena, A.; Acharya, K.; Shah, P.; Bhatt, C.; Nguyen, T.T. Liquidt: Stock market analysis using liquid time-constant neural networks. *Int. J. Inf. Technol.* **2024**, *16*, 909–920. [CrossRef]
10. Sun, S.; Rappaport, T.S.; Rangan, S.; Thomas, T.A.; Ghosh, A.; Kovacs, I.Z.; Rodriguez, I.; Koymen, O.; Partyka, A.; Jarvelainen, J. Propagation path loss models for 5G urban micro-and macro-cellular scenarios. In Proceedings of the 2016 IEEE 83rd Vehicular Technology Conference (VTC Spring), Nanjing, China, 15–18 May 2016; pp. 1–6.
11. European Telecommunications Standards Institute. *5G; Study on Channel Model for Frequencies from 0.5 to 100 GHz*; Technical Report V16.1.0, 3GPP. 2020. Available online: https://www.etsi.org/deliver/etsi_tr/138900_138999/138901/16.01.00_60/tr_138901v160100p.pdf (accessed on 16 June 2024).
12. Xing, Y.; Rappaport, T.S. Millimeter Wave and Terahertz Urban Microcell Propagation Measurements and Models. *IEEE Commun. Lett.* **2021**, *25*, 3755–3759. [CrossRef]
13. Xiao, X.; Hou, X.; Wang, C.; Li, Y.; Hui, P.; Chen, S. Jamcloud: Turning traffic jams into computation opportunities–whose time has come. *IEEE Access* **2019**, *7*, 115797–115815. [CrossRef]
14. Hou, X.; Li, Y.; Jin, D.; Wu, D.O.; Chen, S. Modeling the impact of mobility on the connectivity of vehicular networks in large-scale urban environments. *IEEE Trans. Veh. Technol.* **2015**, *65*, 2753–2758. [CrossRef]
15. Hossan, M.T.; Tabassum, H. Mobility-Aware Performance in Hybrid RF and Terahertz Wireless Networks. *IEEE Trans. Commun.* **2022**, *70*, 1376–1390. [CrossRef]
16. Vankayala, S.K.; Gollapudi, S.K.S.; Jain, B.; Yoon, S.; Mihir, K.; Kumar, S.; Kumar, H.S.; Kommineni, I. Efficient Deep-Learning Models for Future Blockage and Beam Prediction for mmWave Systems. In Proceedings of the NOMS 2023-2023 IEEE/IFIP Network Operations and Management Symposium, Miami, FL, USA, 8–12 May 2023; pp. 1–8.
17. Nagata, H.; Kudo, R.; Takahashi, K.; Ogawa, T.; Takasugi, K. Two-step wireless link quality prediction using multi-camera images. In Proceedings of the 2022 IEEE 33rd Annual International Symposium on Personal, Indoor and Mobile Radio Communications (PIMRC), Kyoto, Japan, 12–15 September 2022; pp. 509–514.
18. Charan, G.; Alkhateeb, A. Computer vision aided blockage prediction in real-world millimeter wave deployments. In Proceedings of the 2022 IEEE Globecom Workshops (GC Wkshps), Rio de Janeiro, Brazil, 4–8 December 2022; pp. 1711–1716.
19. Chen, H.; Sun, C.; Jiang, F.; Jiang, J. Beams selection for MmWave multi-connection based on sub-6GHz predicting and parallel transfer learning. In Proceedings of the 2021 IEEE/CIC International Conference on Communications in China (ICCC), Xiamen, China, 28–30 July 2021; pp. 469–474.
20. Jin, W.; Kim, H.; Lee, H. A Novel Machine Learning Scheme for mmWave Path Loss Modeling for 5G Communications in Dense Urban Scenarios. *Electronics* **2022**, *11*, 1809. [CrossRef]
21. Liu, Y.; Blough, D.M. Environment-aware link quality prediction for millimeter-wave wireless lans. In Proceedings of the 20th ACM International Symposium on Mobility Management and Wireless Access, Montreal, QC, Canada, 24–28 October 2022; pp. 1–10.
22. Nuñez, Y.; Lovisolo, L.; da Silva Mello, L.; Orihuela, C. On the interpretability of machine learning regression for path-loss prediction of millimeter-wave links. *Expert Syst. Appl.* **2023**, *215*, 119324. [CrossRef]

23. Jia, H.; Chen, N.; Zhang, R.; Okada, M. Non-deterministic Quantization for mmWave Beam Prediction. In Proceedings of the 2022 IEEE 35th International System-on-Chip Conference (SOCC), Belfast, UK, 5–8 September 2022; pp. 1–6.
24. Yang, S.; Ma, J.; Zhang, S.; Li, H. Beam Prediction for mmWave Massive MIMO using Adjustable Feature Fusion Learning. In Proceedings of the 2022 IEEE 95th Vehicular Technology Conference:(VTC2022-Spring), Helsinki, Finland, 19–22 June 2022; pp. 1–5.
25. Fu, Z.; Du, F.; Zhao, X.; Geng, S.; Zhang, Y.; Qin, P. A Joint-Neural-Network-Based Channel Prediction for Millimeter-Wave Mobile Communications. *IEEE Antennas Wirel. Propag. Lett.* **2023**, *22*, 1064–1068. [CrossRef]
26. Yu, C.; Sun, Y.; Luo, Y.; Wang, R. mmAlert: mmWave Link Blockage Prediction via Passive Sensing. *IEEE Wirel. Commun. Lett.* **2023**, *12*, 2008–2012. [CrossRef]
27. Ohta, S.; Nishio, T.; Kudo, R.; Takahashi, K.; Nagata, H. Point cloud-based proactive link quality prediction for millimeter-wave communications. *IEEE Trans. Mach. Learn. Commun. Netw.* **2023**, *1*, 258–276. [CrossRef]
28. Pendyala, V.; Kim, H. Assessing the Reliability of Machine Learning Models Applied to the Mental Health Domain Using Explainable AI. *Electronics* **2024**, *13*, 1025. [CrossRef]
29. Sun, S.; MacCartney, G.R.; Rappaport, T.S. A novel millimeter-wave channel simulator and applications for 5G wireless communications. In Proceedings of the 2017 IEEE International Conference on Communications (ICC), Paris, France, 21–25 May 2017; pp. 1–7. [CrossRef]
30. Langdon, W.B.; Poli, R. *Foundations of Genetic Programming*; Springer Science & Business Media: Berlin/Heidelberg, Germany, 2013.
31. Hochreiter, S.; Schmidhuber, J. Long short-term memory. *Neural Comput.* **1997**, *9*, 1735–1780. [CrossRef] [PubMed]
32. Javed, M.A.; Liu, P.; Panwar, S.S. Enhancing XR Application Performance in Multi-Connectivity Enabled mmWave Networks. *IEEE Open J. Commun. Soc.* **2023**, *4*, 2421–2438. [CrossRef]
33. Ju, S.; Kanhere, O.; Xing, Y.; Rappaport, T.S. A millimeter-wave channel simulator NYUSIM with spatial consistency and human blockage. In Proceedings of the 2019 IEEE Global Communications Conference (GLOBECOM), Waikoloa, HI, USA, 9–13 December 2019, pp. 1–6.
34. Shah, S.H.A.; Rangan, S. Multi-Cell Multi-Beam Prediction Using Auto-Encoder LSTM for mmWave Systems. *IEEE Trans. Wirel. Commun.* **2022**, *21*, 10366–10380. [CrossRef]
35. Icke, I.; Bongard, J.C. Improving genetic programming based symbolic regression using deterministic machine learning. In Proceedings of the 2013 IEEE Congress on Evolutionary Computation, Cancun, Mexico, 20–23 June 2013; pp. 1763–1770. [CrossRef]
36. Software, G.G.V. Graphviz [Online]. 2024. Available online: https://graphviz.org/ (accessed on 16 June 2024).
37. Chen, J.; He, Z.S.; Kuylenstierna, D.; Eriksson, T.; Hörberg, M.; Emanuelsson, T.; Swahn, T.; Zirath, H. Does LO Noise Floor Limit Performance in Multi-Gigabit Millimeter-Wave Communication? *IEEE Microw. Wirel. Components Lett.* **2017**, *27*, 769–771. [CrossRef]
38. Virgolin, M.; Pissis, S.P. Symbolic Regression is NP-hard. *arXiv* **2022**, arXiv:2207.01018.

electronics

MDPI

Article

Expedited Near-Field Holographic Microwave Imaging with an Azimuthally Distributed Antenna Array

Mona Heydari † and Reza K. Amineh *,†

Department of Electrical and Computer Engineering, New York Institute of Technology, New York, NY 10023, USA; mona.heydari@nyit.edu

* Correspondence: rkhalaja@nyit.edu

† These authors contributed equally to this work.

Abstract

In this article, we propose a novel near-field holographic microwave imaging technique designed to accelerate the data acquisition process. The system employs a novel electronic switching mechanism utilizing two switching networks that virtually rotate the transmitting and receiving antennas along the azimuthal direction for efficient data collection. This minimizes the need for mechanical scanning of the antennas which, in turn, leads to faster data acquisition. To enhance the quality of the imaging outcome, the number of samples can be increased by combining only a few mechanical scanning steps with the electronic scanning. This data acquisition scheme leverages the system's space-invariant property to enable convolution-based near-field holographic microwave image reconstruction. By capturing and processing scattered fields over a cylindrical aperture, the system achieves high-resolution imaging of concealed objects across multiple range positions. Both simulation and experimental results validate the effectiveness of the proposed approach in delivering high-quality imaging results. Its ability to provide faster and enhanced imaging outcomes highlights its potential for a wide range of applications, including biomedical imaging, security screening, and non-destructive testing of the materials.

Keywords: antenna arrays; microwave holography; microwave measurement; near-field imaging

1. Introduction

Microwave imaging (MWI) has emerged as a transformative technology in various fields due to its ability to penetrate optically opaque materials. Its applications span across various domains, including non-destructive testing (NDT) of industrial materials [1,2], biomedical imaging for early disease detection [3,4], and security screening for concealed threat identification [5,6].

Holographic MWI techniques offer high-speed qualitative imaging capabilities. Traditional holographic systems often rely on the acquisition of wideband data to obtain range resolution [5,7]. Although they have been successfully employed for applications such as security screening of passengers at airports [5], they depend on approximations that restrict their applicability in near-field scenarios, where finer spatial details are critical. In particular, far-field holographic imaging techniques fail to capture and process the evanescent waves emanated by the objects, limiting their resolution and accuracy. To overcome these challenges, near-field holographic imaging systems have been developed to obtain higher resolutions by processing the measured portion of the evanescent waves [8].

In general, the near-field holographic imaging approach, which leverages measured point spread functions (PSFs), provides several advantages over the conventional farfield holographic imaging techniques, including the following: (1) Measured PSFs accurately capture the near-field behavior of the antennas used in the specific imaging setup, leading to more realistic system modeling. (2) Unlike analytical PSFs in the far-field holography, which assume scalar far-field values at isolated points, measured PSFs represent actual responses such as S-parameters or terminal voltages. (3) Measured PSFs reflect the true electromagnetic properties of the background medium and include the effects of all components in the imaging system—such as antennas, positioners, and measurement chambers—avoiding the oversimplified assumptions used in far-field holography. (4) By using a calibrated scattering probe of known material and size, the system's sensitivity to contrast can be quantified. This enables accurate and quantitative imaging [9,10]. (5) The system of equations solved in near-field holographic imaging at each spectral point is smaller and better conditioned than those in optimization-based microwave imaging methods [11]. This not only improves the computational efficiency but also allows for natural parallelization, as each spectral point can be processed independently. (6) Near-field holographic imaging eliminates the need for interpolation in the spectral domain (as implemented in far-field holography [8]), simplifying the reconstruction process. (7) Unlike far-field holographic imaging, near-field holographic imaging does not rely on the assumption that the spatial frequency variables are independent, thereby reducing reconstruction artifacts. (8) Data from multiple antennas, including both forward- and back-scattered signals, can be combined within a single linear system to enhance spatial resolution and reduce image artifacts. (9) Near-field holographic imaging performs effectively with a smaller number of frequency samples, particularly when multiple transmitter/receiver channels are used. This contrasts with far-field holography, which requires densely sampled frequencies due to its reliance on a three-dimensional (3D) inverse Fourier transform.

We emphasize that the near-field holographic imaging technique is based on the Born approximation, which assumes that the incident field in the presence of the object is the same as the incident field in the absence of the object. This assumption linearizes the scattering integral and allows for the use of convolution theory (which holds only for linear systems). However, it provides the best results primarily for small or low-contrast objects (weak scatterers).

In particular, the condition for applying the first-order Born approximation is that the radius a of a sphere enclosing the object and its refractive index n need to satisfy $(n-1)a < \lambda/4$, where λ is the wavelength in the background medium [12,13]. Thus, when the objects or their contrasts (with respect to the background medium) become larger, or when there are multiple objects inside the imaged domain, the Born assumption imposes imaging errors due to further disturbance of the linearity of the imaging system.

This challenge can be overcome using iterative optimization methods such as the Born iterative method (e.g., see [14]), distorted Born iterative method (e.g., see [15]), hybrid Born–Rytov method (e.g., see [10]), Newton-type methods (e.g., see [16]), contrast source inversion (e.g., see [17]), or hybrid and data-driven techniques such as machine learning/deep learning techniques (e.g., see [18]), etc. However, these techniques have certain drawbacks in practical applications, including being extremely demanding in terms of memory and time, sensitivity to initial guess, ill-posedness of the solution, and the requirement for large datasets.

Although near-field holographic imaging techniques initially required wideband information to perform three-dimensional (3D) imaging [19], later these techniques used a receiver antenna array to perform 3D imaging using narrow-band data [20]. Narrow-band near-field holographic imaging systems have also been developed for cylindrical setups, where an array of receiver antennas collects scattered fields while rotating, together with the transmitter antenna, around the imaged medium [21]. Despite these advancements, significant obstacles to employing these techniques in applications demanding high-speed imaging remain. In particular, in [21], data acquisition time is a burden since responses need to be collected over a cylindrical aperture by mechanically scanning a transmitter and an array of receiver antennas.

To speed up the data acquisition, electronic scanning of an antenna array can be used instead of mechanical scanning. One method is using a two-dimensional (2D) array of antennas covering the aperture (e.g., see [22]). However, this method is costly due to the use of a large number of antennas and the switching network to control the data acquisition. Another common method that alleviates the cost issue involves arranging antenna arrays in one direction while mechanically scanning them along the perpendicular direction (e.g., see [23–25]). Thus, in this work, we take this principle into practice and we build a system featuring the space-invariant property by mimicking mechanical scanning along the azimuthal axis as implemented in [21] via an array of antennas switched electronically. The system's space-invariant property means that when an object shifts along the cross-range direction (azimuthal or longitudinal directions), its response remains the same but shifts with the same amount and along the same direction [26]. This allows the scattered field measurements to be modeled as a convolution between the point spread function (PSF) and the object's reflectivity profile, enabling efficient image reconstruction using Fourier-based techniques [8]. To build such a system, we develop a novel switching mechanism utilizing two switching networks that virtually (electronically) rotate the transmitter and the array of receivers around the imaged medium for data collection. While mechanical scanning is still required along the longitudinal axis, the electronic data collection along the azimuthal axis expedites the data acquisition process significantly. We also demonstrate the possibility of collecting more samples along the azimuthal axis via combining a few mechanical scanning steps with the electronic scanning. This alleviates the coarse sampling in the electronic scanning scheme, leading to further enhancement of the reconstructed images. The performance of the proposed imaging techniques will be demonstrated via simulation results and experiments.

2. Methodology

This work proposes a microwave imaging system comprising an array of N_A antennas uniformly distributed along the azimuthal axis to cover the full circular aperture. The angular separation between adjacent antennas is denoted by $\Delta\phi$. In this configuration, microwave switches are used to control the antenna operations such that, at each electronic scanning step, one antenna functions as the transmitter while a set of N_R antennas positioned on the opposite side of the imaging domain act as receivers. Also, we present an approach to increase the number of samples along the azimuthal direction by adding a few mechanical scanning steps in the azimuthal direction.

2.1. Data Collection Schemes

To collect data on a cylindrical aperture, mechanical scanning is first performed along the vertical z axis, incrementally moving the antenna array to different heights. Then, at each fixed height, data are acquired along the azimuthal direction using one of two scanning schemes.

- Electronic scanning along azimuthal direction: In this scheme, the physical antenna array remains stationary while the active transmitting and receiving elements are electronically switched through two switching networks. The switching networks are controlled via an Arduino-based system. This enables data acquisition from multiple angular positions without the need for physical rotation, offering faster scanning speed and reducing mechanical complexity. This scheme is discussed in Section 2.1.1.

- Electronic–mechanical scanning along azimuthal direction: Due to physical limitations on the number of antennas that can be placed along the azimuthal direction, the angular resolution of purely electronic scanning will be constrained. To address this, a hybrid method is employed, wherein the entire array is rotated incrementally around the azimuthal (φ) axis using a stepper motor. At each mechanical rotation step, the full electronic scanning sequence (as described above) is repeated. This enhances the angular sampling density by a factor of N_{mech} (the number of mechanical steps), effectively increasing the total number of azimuthal samples at each height. This approach balances enhanced quality with manageable system complexity. This scheme is discussed in Section 2.1.2.

2.1.1. Electronic Scanning Along the Azimuthal Axis

A straightforward approach to implement the electronic scanning scheme is to construct the system with two $1 \times N_A$ switching networks—one for the transmitter path and another one for the receiver path—along with additional N_A switch elements of 1×2 to control the role of each antenna (i.e., transmitter or receiver). However, this configuration is costly, bulky, and complex to implement. To address this, we propose a novel switching scheme involving only two switching networks (labeled as A and B), each of size $1 \times (N_A/2)$, assuming N_A is even. The common (COM) ports of these networks are connected to the two ports of a vector network analyzer (VNA) for measurement. The output ports of switching networks A and B are alternately connected to the elements of the antenna array in an interleaved manner, as illustrated in Figure 1. Antennas associated with networks A and B are shown in red and green, respectively. Electronic scanning is performed such that, at each scanning step along ϕ, an antenna from one of the networks operates as the transmitter, illuminating the imaged medium, while N_R antennas from the opposite network act as receivers. These N_R receivers are chosen across an angular span of $2 \times N_R \times \Delta\phi$, with the transmitting antenna located at the center of this span but on the opposite side of the imaged domain (see Figure 1 for reference). This selection and measurement process is repeated until each of the N_A antennas has acted as a transmitter once in the specific z position. This procedure ensures scattered field measurements over a full circular aperture. Figure 1a,b show two such sampling steps.

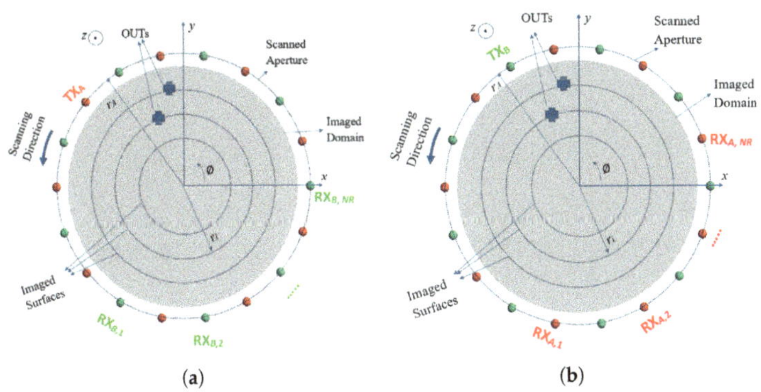

Figure 1. Cross-section illustration of the proposed imaging system. Small circles show the antennas distributed uniformly along the azimuthal direction. Antennas marked in red and green are connected to switching networks A and B, respectively. Two sample scanning steps are shown: (**a**) an antenna from switching network A is selected as a transmitter TX_A while the receivers RX_B are selected from network B; (**b**) an antenna from network B is selected as a transmitter TX_B while the receivers RX_A are selected from network A.

In Figure 1a, an antenna (TX_A) from switching network A (marked in red color) is assigned the role of transmitter, while the receivers ($RX_{B,1}$ to RX_{B,N_R}) are selected from network B (marked in green color). Conversely, in Figure 1b, an antenna (TX_B) from network B acts as the transmitter, with the receivers ($RX_{A,1}$ to RX_{A,N_R}) selected from network A.

By implementing this data acquisition configuration at multiple heights N_z, data is acquired using one active transmitter and N_R receiver antennas operating in coordination, forming a virtual cylindrical aperture with radius r_A and height z_A. Thus, the scattered field $E^{sc}_{a_m}(\phi, z)$ for each virtual receiver $a_m, m = 1, \ldots, N_R$ is recorded at N_ϕ angles along the azimuthal direction within 0 to 2π. The process of data collection is performed at N_ω frequencies ($\omega_n, n = 1, \ldots, N_\omega$).

2.1.2. Combination of Electronic and Mechanical Scanning Along the Azimuthal Axis

The physical size of the antennas imposes some limitation on the number of elements used in the array along the azimuthal axis. This, in turn, reduces the number of samples that can be collected along that axis in electronic scanning. To tackle this issue, the number of scanning steps can be increased by adding a few mechanical scanning steps along the ϕ direction. At each mechanical scanning step, we repeat the electronic scanning steps discussed in Section 2.1.1.

To clarify further, in this procedure, for each z-position, electronic and mechanical scanning is performed around the aperture. Although electronic scanning is performed with sampling steps of $\Delta\phi$, adding a few mechanical scanning steps N_{mech} and repeating electronic scanning at each step leads us to reach our objective to increase the number of samples by N_{mech} times. As a result, the total number of samples collected at each z-position is calculated as $N_{mech}\left(\frac{2\pi}{\Delta\phi}\right)$.

2.2. Image Reconstruction Technique

The image reconstruction process is supposed to provide images of objects under test (OUTs) over cylindrical surfaces with radii r_i, where i ranges from 1 to N_r and r_i lies within the interval $(0, r_A)$. For the implementation of the holographic imaging, the approximate point spread functions (PSFs) $E^{sc,PWO}_{i,a_m}(\phi, z)$ of the system are measured by placing small point-wise objects (PWO) at cylindrical coordinates (r, ϕ, z) of $(r_i, 0, 0)$ (with reference to Figure 1), one at a time, and scanning the responses over the cylindrical aperture. In total, we measure one PSF function for each imaged surface N_r. According to [20], the total measured scattered field by each virtual receiver antenna and at each frequency can be written as:

$$E^{sc}_{a_m}(\phi, z, \omega_n) = \sum_{i=1}^{N_r} E^{sc,PWO}_{i,a_m}(\phi, z, \omega_n) *_\phi *_z f_i(\phi, z) \quad (1)$$

where $*_\phi$ and $*_z$ are convolutions along the azimuthal and longitudinal axes, respectively. The goal is to estimate the contrast functions $f_i(\phi, z)$. For each receiver, the equation above is rewritten across all frequencies as below.

$$\begin{cases} E^{sc}_{a_m}(\phi, z, \omega_1) = \sum_{i=1}^{N_r} E^{sc,PWO}_{i,a_m}(\phi, z, \omega_1) *_\phi *_z f_i(\phi, z) \\ \vdots \\ E^{sc}_{a_m}(\phi, z, \omega_{N_\omega}) = \sum_{i=1}^{N_r} E^{sc,PWO}_{i,a_m}(\phi, z, \omega_{N_\omega}) *_\phi *_z f_i(\phi, z) \end{cases} \quad (2)$$

Discrete Fourier transform (DFT) along the ϕ axis and discrete time Fourier transform (DTFT) along the z axis are implemented on the equations to transform them to the spectral domain. This allows for forming a system of equations at each Fourier variable pair $\kappa = (k_\phi, k_z)$ as shown below that is solved using a standardized minimum norm approach [27]:

$$\tilde{\tilde{\mathbf{E}}}^{sc}_{(N_R N_\omega \times 1)} = \tilde{\tilde{\mathbf{D}}}_{(N_R N_\omega \times N_r)} \tilde{\tilde{\mathbf{F}}}_{(N_r \times 1)} \tag{3}$$

$$\tilde{\tilde{\mathbf{E}}}^{sc} = \mathbf{L} \begin{bmatrix} \tilde{\tilde{\mathbf{E}}}^{sc}_1 \\ \vdots \\ \tilde{\tilde{\mathbf{E}}}^{sc}_{N_R} \end{bmatrix} \tag{4}$$

$$\tilde{\tilde{\mathbf{D}}} = \mathbf{L} \begin{bmatrix} \tilde{\tilde{\mathbf{D}}}_1 \\ \vdots \\ \tilde{\tilde{\mathbf{D}}}_{N_R} \end{bmatrix} \tag{5}$$

$$\tilde{\tilde{\mathbf{F}}} = \begin{bmatrix} \tilde{\tilde{f}}_1(\boldsymbol{\kappa}) \\ \vdots \\ \tilde{\tilde{f}}_{N_r}(\boldsymbol{\kappa}) \end{bmatrix} \tag{6}$$

$$\tilde{\tilde{\mathbf{E}}}^{sc}_{a_m} = \begin{bmatrix} \tilde{\tilde{E}}^{sc}_{a_m}(\boldsymbol{\kappa}, \omega_1) \\ \vdots \\ \tilde{\tilde{E}}^{sc}_{a_m}(\boldsymbol{\kappa}, \omega_{N_\omega}) \end{bmatrix} \quad \text{of size } (N_\omega \times 1) \tag{7}$$

$$\tilde{\tilde{\mathbf{D}}}_{a_m} = \begin{bmatrix} \tilde{\tilde{E}}^{sc,\mathrm{PWO}}_{1,a_m}(\boldsymbol{\kappa}, \omega_1) & \cdots & \tilde{\tilde{E}}^{sc,\mathrm{PWO}}_{N_r,a_m}(\boldsymbol{\kappa}, \omega_1) \\ \vdots & \ddots & \vdots \\ \tilde{\tilde{E}}^{sc,\mathrm{PWO}}_{1,a_m}(\boldsymbol{\kappa}, \omega_{N_\omega}) & \cdots & \tilde{\tilde{E}}^{sc,\mathrm{PWO}}_{N_r,a_m}(\boldsymbol{\kappa}, \omega_{N_\omega}) \end{bmatrix} \quad \text{of size } (N_\omega \times N_r) \tag{8}$$

where $\tilde{\tilde{E}}^{sc}_{a_m}(\boldsymbol{\kappa}, \omega_n)$, $\tilde{\tilde{E}}^{sc,PWO}_{i,a_m}(\boldsymbol{\kappa}, \omega_n)$, and $\tilde{\tilde{f}}_i(\boldsymbol{\kappa})$ are obtained from $E^{sc}_{a_m}(\phi, z, \omega_n)$, $E^{sc,PWO}_{i,a_m}(\phi, z, \omega_n)$, and $f_i(\phi, z)$ when taking DFT and DTFT along the azimuthal and longitudinal directions, respectively; $\underline{\underline{\mathbf{L}}}(N_\omega N_R \times N_\omega N_R) = \underline{\underline{\mathbf{I}}} - \underline{\mathbf{U}}\underline{\mathbf{U}}^{\mathrm{T}} / \underline{\mathbf{U}}^{\mathrm{T}}\underline{\mathbf{U}}$, $\underline{\underline{\mathbf{I}}}(N_\omega N_R \times N_\omega N_R)$ is the identity matrix; and $\underline{\mathbf{U}}_{(N_\omega N_R \times 1)}$ is a vector of ones. In the standardized minimum norm approach, the standardized solution $\hat{\tilde{\tilde{\underline{\mathbf{F}}}}}$ is obtained as:

$$\hat{\tilde{\tilde{\underline{\mathbf{F}}}}} = \sqrt{\left(\mathrm{Diag}\left(\underline{\underline{\mathbf{S}}}_{\tilde{\tilde{\mathbf{F}}}}\right)\right)^{-1}} \underline{\underline{\mathbf{P}}} \tilde{\tilde{\underline{\underline{\mathbf{E}}}}}^{sc} \tag{9}$$

where $\underline{\underline{S}}_{\tilde{\tilde{\mathbf{F}}}(N_r \times N_r)}$ is the variance of $\tilde{\tilde{\underline{\mathbf{F}}}}$ obtained from:

$$\underline{\underline{\mathbf{S}}}_{\tilde{\tilde{\mathbf{F}}}} = \underline{\underline{\mathbf{P}}} \mathbf{S}_{\underline{\underline{\tilde{\tilde{\mathbf{E}}}^{sc}}}} \underline{\underline{\mathbf{P}}}^H \tag{10}$$

where $\mathbf{S}_{\underline{\underline{\tilde{\tilde{\mathbf{E}}}^{sc}}}}$ is the variance of $\tilde{\tilde{\underline{\underline{\mathbf{E}}}}}^{sc}$ from a Bayesian point of view, $\mathrm{Diag}\left(\underline{\underline{\mathbf{S}}}_{\tilde{\tilde{\mathbf{F}}}}\right)$ is the diagonal matrix obtained from diagonal elements of $\underline{\underline{\mathbf{S}}}_{\tilde{\tilde{\mathbf{F}}}}$, and $\underline{\underline{\mathbf{P}}}$ is

$$\underline{\underline{\mathbf{P}}} = \underline{\underline{\tilde{\tilde{\mathbf{D}}}}}^H \left[\underline{\underline{\tilde{\tilde{\mathbf{D}}}\tilde{\tilde{\mathbf{D}}}}}^H + \alpha \underline{\underline{\mathbf{L}}} \right]^+ \tag{11}$$

where $[.]^H$ is the Hermitian transpose operation, $[.]^+$ is the Moore–Penrose pseudoinverse, and α is the regularization parameter chosen empirically.

Solving the systems of equations in (3) at all $\boldsymbol{\kappa} = (k_\phi, k_z)$ pairs provides the images $\tilde{\tilde{f}}_i(\boldsymbol{\kappa})$ in the spectral domain. Lastly, inverse DFT along ϕ and inverse DTFT along z are applied to obtain the reconstructed images $f_i(\phi, z)$ in the spatial domain at each imaged surface at r_i.

3. Simulations

This section evaluates the performance of the proposed imaging technique using simulated data obtained from FEKO [28]. Simulations were carried out at two operating frequencies: 1.5 GHz and 1.8 GHz.

With the aim of covering the full circle and considering the size of antennas used in the experiments (discussed later), we used an angular separation of 10 deg between the antennas, which led to having 36 antennas.

In the simulations, all the utilized antennas were z-polarized resonant dipoles. The background medium was characterized by $\varepsilon_r = 22.1$ and $\sigma = 0.2$ S/m, while the imaged objects (OUTs) were assigned $\varepsilon_r = 55$ and $\sigma = 4$ S/m. The scattered fields were acquired over a cylindrical aperture with radius $r_A = 60$ mm and height $z_A = 188$ mm.

The imaging was performed on three radial surfaces: 24 mm, 36 mm, and 48 mm. The point spread functions (PSFs) for small 4 mm cuboids placed at these radial distances were simulated. To mimic a realistic environment, additive White Gaussian noise was introduced to the scattered data with a signal-to-noise ratio (SNR) of 30 dB.

Sampling in the longitudinal (z) direction was performed at intervals of $\Delta z = \lambda/4$, where λ is the wavelength at the corresponding frequency. The results of the two proposed scanning schemes discussed earlier are presented below for two challenging imaging scenarios.

Figure 2a illustrates the FEKO simulation setup for Simulation Example 1, where three linear cuboidal objects are positioned at radii r_1 and r_3. Two of them are placed at r_3 with angular positions φ_3 and $-\varphi_3$, while one is aligned at r_1 along the x axis. Table 1 summarizes the parameters for this example.

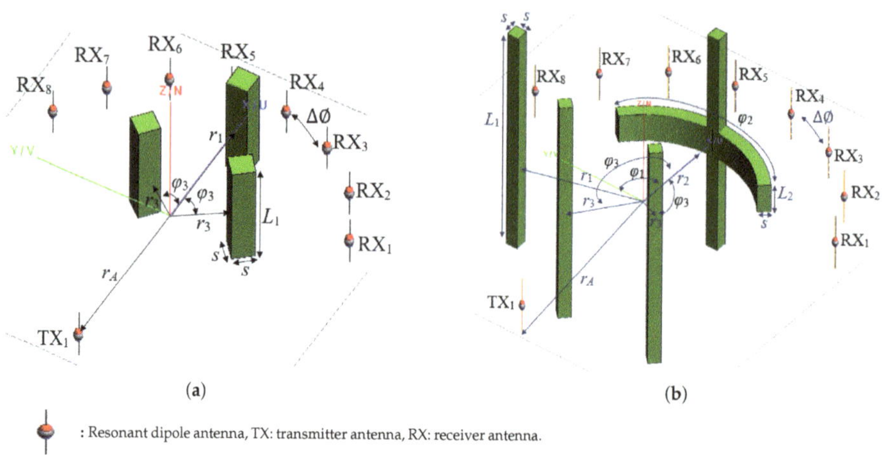

Figure 2. Simulated models generated using FEKO software. (**a**) Simulation Example 1: Three cuboid objects (green) positioned on two radial surfaces. (**b**) Simulation Example 2: Three cuboids and one curved T-shaped object distributed across three radial surfaces.

Table 1. Values of parameters for Simulation Example 1 shown in Figure 2a.

Parameter	r_1	r_3	L_1	φ_3	s	r_A
Value	48	24	37	50	8	60
Unit	mm	mm	mm	deg	mm	mm

To further examine the reconstruction performance, we simulate a more complex scenario example depicted in Figure 2b, involving four objects. Simulation Example 2 involves a more complex arrangement with four distinct objects distributed across three radial surfaces, as compared to the simpler configuration in Figure 2a. This example is included to further demonstrate the system's capability to handle more intricate geometries and object distributions. The reconstruction results from Example 2 provide evidence of the method's robustness in accurately localizing and resolving multiple targets under more challenging conditions. According to the parameters listed in Table 2, two cuboids are located at r_3, one at r_1, and a more complex structure with two orthogonal arms is placed at r_2.

Table 2. Values of parameters for Simulation Example 2 shown in Figure 2b.

Parameter	r_1	r_2	r_3	L_1	s	L_2	r_A	φ_1	φ_2	φ_3
Value	48	36	24	75	4	9	60	90	135	135
Unit	mm	mm	mm	mm	mm	mm	mm	deg	deg	deg

3.1. Results of Electronic Scanning Along Azimuthal Direction

Instead of employing a stationary array of $N_A = 36$ antennas along the azimuthal direction, the simulation setup involves rotating a transmitter and eight receiver antennas ($N_R = 8$ with reference to Figure 1) 360 deg around the imaging domain with a step size of $10°$, thereby emulating a $10°$ separation in a fixed array configuration. The chosen number of receiver antennas here is inspired by the work performed in [21], where mechanical scanning was used to collect data along the ϕ axis. The angular separation between receiver antennas is $\Delta\phi = 20$ deg, consistent with the stationary antenna configuration.

For image reconstruction, the regularization parameter α in Equation (11) is chosen empirically, which is a common method (e.g., see [29,30]). To optimize the value of α, the quality of the reconstructed images is evaluated using the structural similarity (SSIM) index [31].

In summary, SSIM is computed based on three terms: the luminance term, the contrast term, and the structural term. To find the optimal value of α, SSIM is computed for the reconstructed images using the true object's image as the reference. The true image has a value of 1 at the pixels overlapping the object and 0 elsewhere.

The overall SSIM is computed as the sum of SSIMs calculated for the individual two-dimensional (2D) images at the three radial positions in each image reconstruction process. Therefore, a higher value of total SSIM indicates a greater similarity to the true images.

Please note that this process needs to be performed for at least one set of known objects to optimize the regularization parameter. Here, we perform this for Simulation Example 1 and then use the optimized value for Simulation Example 2. Table 3 shows the variation in total SSIM versus the value of the regularization parameter for Simulation Example 1.

Table 3. Variation in total SSIM versus the value of regularization parameter α for Simulation Example 1.

Regularization Parameter α	Total SSIM	Observation
11×10^{-12}	0.62	
11×10^{-11}	0.87	
11×10^{-10}	1.21	
11×10^{-9}	1.42	
$\mathbf{11 \times 10^{-8}}$	**1.44**	Optimal [1]
11×10^{-7}	1.29	
11×10^{-6}	0.97	
11×10^{-5}	0.96	

[1] This value of α gives the best total SSIM and is used in subsequent simulation results.

It was observed that the most acceptable reconstruction quality occurs around $\alpha = 11 \times 10^{-8}$. As illustrated in Figure 3a,b, using much higher or lower values such as $\alpha = 11 \times 10^{-7}$ or $\alpha = 11 \times 10^{-10}$ results in poor image reconstruction, where the object features are not clearly distinguishable. Thus, we present the results below using $\alpha = 11 \times 10^{-8}$.

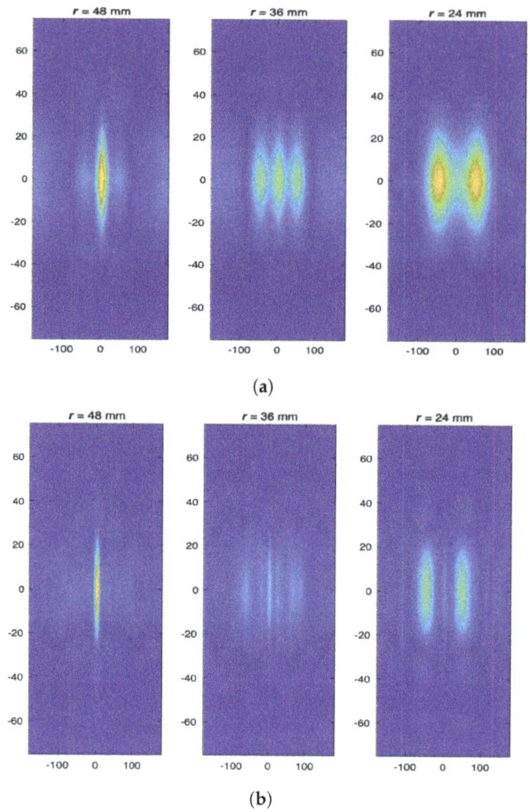

Figure 3. Image reconstruction results for different values of the regularization parameter α which are far from the optimized value for Simulation Example 1, (**a**) for $\alpha = 11 \times 10^{-7}$ and (**b**) for $\alpha = 11 \times 10^{-10}$.

Figure 4a presents the reconstructed images for Simulation Example 1. The results indicate successful localization and shape recovery of the targets. It is also evident that image resolution is superior on the outer surface at $r_1 = 48$ mm compared to the inner surface at $r_3 = 24$ mm. This improvement is attributed to enhanced accessibility and interaction with evanescent components in near-field holography.

The reconstructed images for Simulation Example 2 are displayed in Figure 4b. The results demonstrate clear localization and differentiation of targets at r_1 and r_3. For the object placed at r_2, the contrast is more pronounced along the z axis, aligned with the z-polarized dipole antenna response. As complexity increases, some imaging artifacts and shadowing are observed across adjacent radial surfaces.

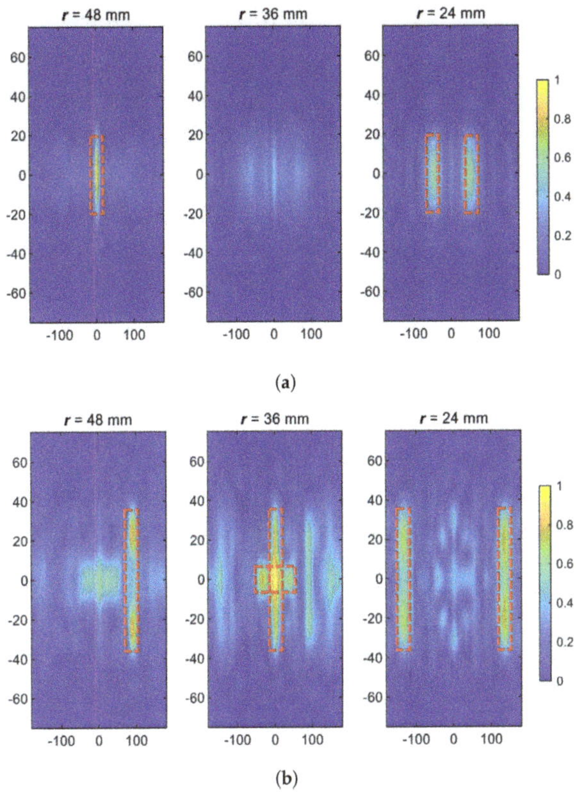

Figure 4. Reconstructed images for (a) Simulation Example 1 and (b) Simulation Example 2 when 36 samples have been used along the ϕ axis and electronic scanning along the azimuthal direction has been performed. Red dashed lines outline the target boundaries. Horizontal axis: ϕ (deg); vertical axis: z (mm).

3.2. Simulation Results of Electronic and Mechanical Scanning Along Azimuthal Direction

By adding a few mechanical scanning steps N_{mech} and repeating the electronic scanning at each one of those steps, as discussed in Section 2.1.2, the number of collected samples along the azimuthal direction is multiplied by N_{mech}. With the consideration of the experimental setup discussed later, the overall number of 201 samples along the azimuthal direction is calculated and applied in the simulation. Figure 5 shows the simulation results of this data acquisition scheme (201 samples along the ϕ axis) for the two examples shown in Figure 2. Comparing the reconstructed images in Figures 4 and 5, it can clearly be observed that the qualities of the reconstructed images are enhanced and two common issues in the reconstruction of images (having object shadows on adjacent planes and spurious artifacts) have been reduced significantly, when a larger number of samples along the azimuthal axis has been acquired.

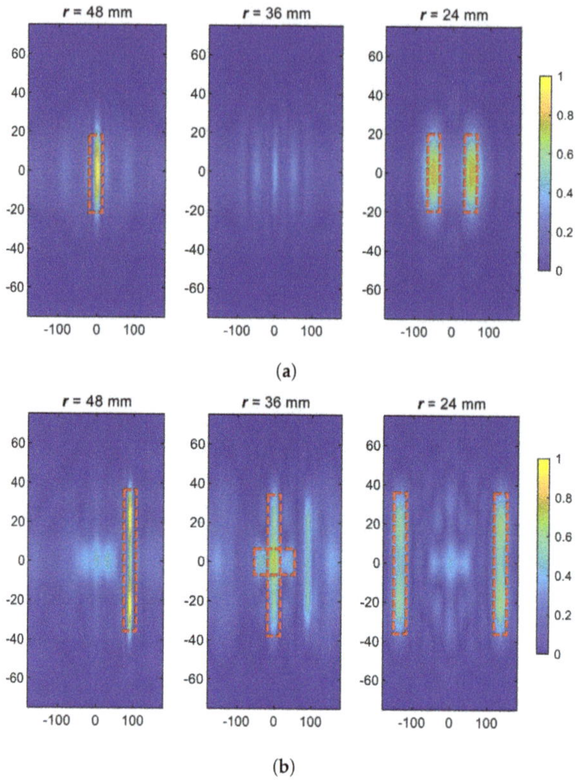

Figure 5. Reconstructed images for (**a**) Simulation Example 1 and (**b**) Simulation Example 2 when 201 samples were used along the ϕ axis and electronic and mechanical scanning along the azimuthal direction was performed. Red dashed lines outline the target boundaries. Horizontal axis: ϕ (deg); vertical axis: z (mm).

4. Experimental Results

Figure 6 illustrates the block diagram of the proposed imaging system. As shown, it consists of an array of antennas and two switching networks A and B to select the antennas, as discussed in Section 2. The array elements are mini GSM/Cellular Quad-Band antennas from Adafruit [32]. These antennas are a practical, affordable, and technically suitable choice for this microwave imaging system. They operate around 1.5–1.8 GHz, which matches the imaging system's narrow-band frequencies and allows effective penetration into the glycerine–water medium. Their small size makes it easier to arrange many antennas around a cylindrical setup, achieving higher angular sampling rate. Also, we employ an Anritsu VNA, ShockLine-MS46122B model, for the measurements.

The scanning setup includes a plexiglass container with a diameter of 125 mm and a height of 200 mm, filled with a liquid consisting of 20% water and 80% glycerin. The liquid has properties of $\varepsilon_r = 22$ and $\sigma = 1.25$ S/m [33]. This mixture has been chosen due to the similarities of its properties to the averaged tissue properties in biomedical applications [33].

To mitigate the potential influence of the air–liquid interface on the measured signals, the antenna holder was custom designed to tightly accommodate the container and minimize the surrounding gap. This configuration reduces the potential of having interface-related distortions while preserving sufficient leeway to allow mechanical movement of the container along the azimuthal (φ) and longitudinal (z) directions during the scanning process.

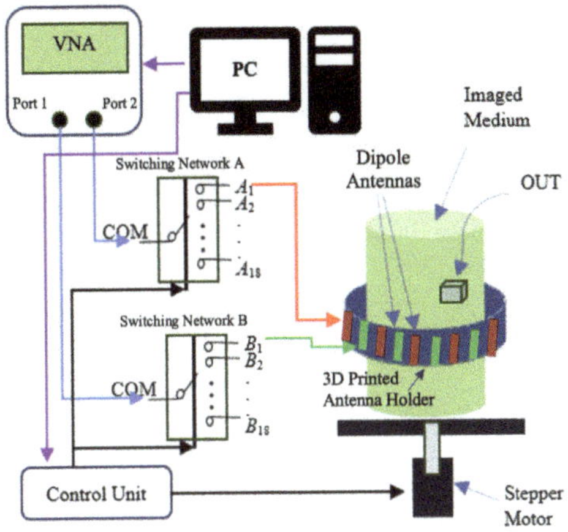

Figure 6. Block diagram of the imaging system used for measurement. Two ports of the VNA are connected to the common ports of two switching networks. One stepper motor is used for mechanical scanning along the azimuthal direction and another stepper motor (not shown) is used for scanning along the z axis. Dipole antennas (shown by red and green) are connected to the two switching networks A and B, respectively. The objects under test (OUTs) are 3D-printed targets covered by copper sheets.

The positioning system, which includes an Arduino Uno board, an Arduino motor shield board, and a stepper motor, moves the liquid container along the longitudinal (z) direction. During the measurements, the antenna array remains stationary and is housed in a customized 3D-printed holder, ensuring a gap of 1 mm with the liquid container to allow smooth container movement. We use 36 antennas distributed evenly along the azimuthal direction, covering the full circle. To minimize unwanted cross-coupling between the antennas and reduce the impact of external electromagnetic interference, small microwave-absorbing sheets are placed inside the antenna array holder between adjacent antennas. Specifically, microwave-absorbing sheets with dimensions of 50 mm × 15 mm are inserted into the slots of the holder between each pair of adjacent antennas to reduce mutual coupling. Additionally, the outer surface of the holder is lined with a layer of microwave-absorbing sheet. To further improve shielding, the entire measurement setup is enclosed in a custom-built wooden box, which is also lined with microwave-absorbing material. This layered shielding approach ensures that the measured S-parameters are minimally affected by environmental noise, resulting in more accurate and consistent data for image reconstruction.

Figure 7 shows the variation in the scattering parameters of the antennas. As depicted, Figure 7a shows the variation in the reflection scattering parameter for a sample antenna placed inside the antenna holder and in contact with the liquid container, while Figure 7b shows the variation in the transmission scattering parameter measured for two sample antennas located at the opposite sides of the liquid container. While the figures show the variation over frequencies 1 GHz to 2 GHz, the imaging is performed using the data collected over 1.5 GHz to 1.8 GHz. Figure 7 shows the acceptable performance of the antennas over this frequency range for imaging.

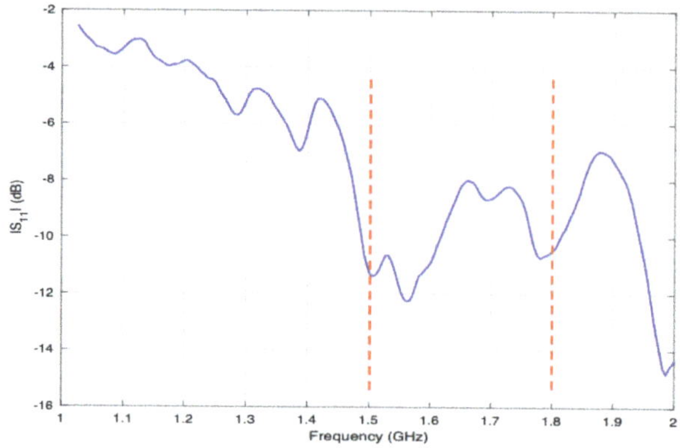

(**a**)

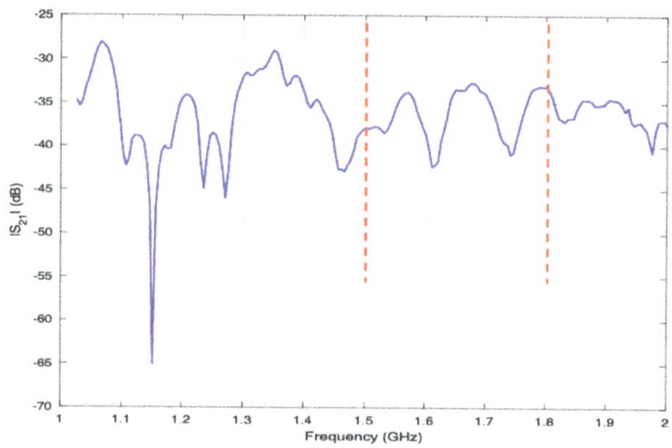

(**b**)

Figure 7. (**a**) Variation in reflection scattering parameter for a sample antenna. (**b**) Variation in transmission scattering parameter for two antennas placed at opposite sides of the liquid container. The red dashed lines indicate the frequency range used for image reconstruction.

The selection of an antenna's role as transmitter or receiver is controlled by the two switch networks of 1×18. Each network is implemented using an Arduino Uno board and three EV1HMC321ALP4E modules, which are RF SP8T switches from Analog Devices. Each network includes a master and two slave modules, controlled by MATLAB through its assigned Arduino Uno board.

Referring to the theory discussed earlier, to perform electronic scanning, the selection of antennas' roles (transmitter or receiver) is determined such that, while one antenna from a network is selected as a transmitter, the corresponding eight antennas connected to the other network (on the opposite side of the imaged medium) act as receivers. The common ports of the switch networks are connected to the two ports of the VNA, and the transmission scattering parameter $S_{21}(TX_i, RX_j, z_k, \omega_n)$ is measured for every pair of transmitter (TX_i) and receiver (RX_j) antennas ($i, j = 1, \cdots, N_A$) at each scanned height ($z_k, k = 1, \cdots, N_z$) and measured frequency ($\omega_n, n = 1, \cdots, N_\omega$).

The assumption of identical antennas is not guaranteed in practice due to fabrication variations, connector mismatches, etc. In the previous work [20], where mechanical scanning was used and each antenna had a fixed role, this issue did not cause significant effects. However, in this work, due to the dynamic assignment of the role of transmitter or receiver to different antennas during electronic scanning, compensation for these variations must be addressed. To overcome this issue, the raw responses of the antennas are pre-processed as

$$S_{21}^{comp}(TX_i, RX_j, z_k, \omega_n) = \frac{S_{21}^{wo}(TX_i, RX_j, z_k, \omega_n)}{S_{21}^{no}(TX_i, RX_j, z_k, \omega_n)} \quad (12)$$

where S_{21}^{comp} is the compensated response, S_{21}^{wo} is the measured response with the object, and S_{21}^{no} is the response without an object.

The approximation of no-object data is obtained by analyzing the variation in the measured signal along the z-direction. It is assumed that regions with higher signal variation correspond to the positions of the objects, while regions with minimal variation are considered no-object zones. These low-variation regions are used as a reference to approximate the background response.

Figure 8 shows a photo of the experimental setup placed inside a shielding box. The PSFs are collected for objects positioned at radial distances of 20 mm, 35 mm, and 50 mm inside the liquid container, allowing reconstruction of the images on surfaces at these radii. The OUTs are plastic objects covered with thin copper sheets and placed inside the liquid container. Data acquisition is performed over the frequency range from 1.6 GHz to 1.75 GHz with steps of 0.05 GHz.

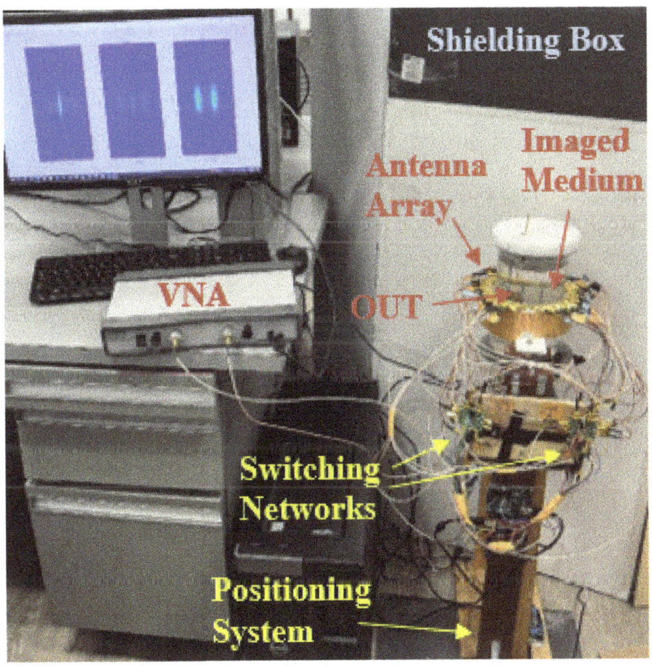

Figure 8. Photo of the imaging system used for the measurements.

In order to test the proposed scheme in practice, four different measurement scenarios have been performed, as discussed below.

For the first Measurement Example, the OUTs are in the form of cylinders with a height of 50 mm and a diameter of 20 mm, placed at the cylindrical coordinates (r, ϕ, z) of (35 mm, 290 deg, 0) and (50 mm, 180 deg, 0). The mechanical scanning along the z axis is performed from −80 mm to 80 mm with 31 steps. Figure 9a shows the reconstructed images of the positioned objects in the imaged medium. Despite the inevitable presence of environmental interferences and measurement noise, the objects are clearly distinguishable in the images.

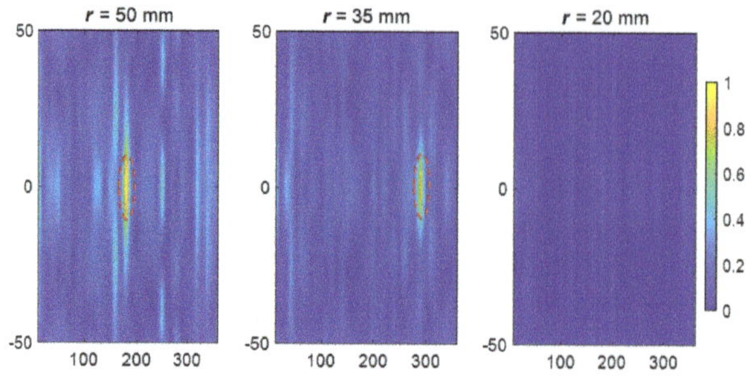

(**a**)

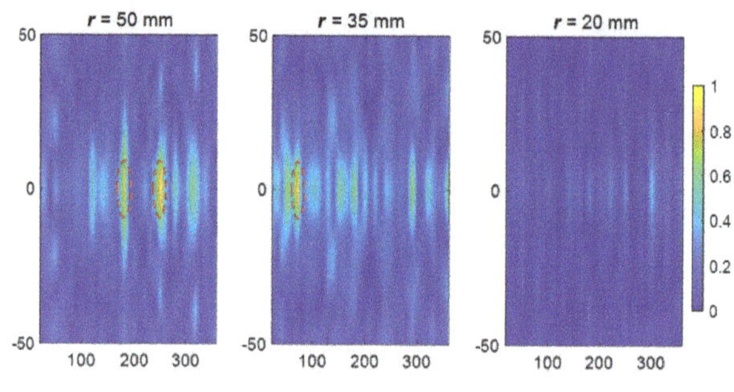

(**b**)

Figure 9. Measurement results for (**a**) Measurement Example 1 and (**b**) Measurement Example 2 when 36 samples along the ϕ axis are used. The red dashed lines show the positions of the objects. For all the figures, the horizontal axis represents ϕ (deg) and the vertical axis represents z (mm).

For the second Measurement Example, the OUTs are small cuboids with dimensions of 20 mm $\times$ 20 mm $\times$ 8 mm, placed at the cylindrical coordinates (r, ϕ, z) of (35 mm, 70 deg, 0), (50 mm, 180 deg, 0), and (50 mm, 250 deg, 0). The mechanical scanning along the z axis is carried out from -50 mm to 50 mm in 19 steps. The result of the experiment is shown in Figure 9b. As observed, the objects are reconstructed as expected in their correct positions. In this example, due to the larger number of OUTs, an increase in the level of artifacts is observed, as expected.

In the third measurement experiment, the proposed data acquisition scheme of the combination of electronic and mechanical scanning has been tested and verified practically. With the help of an additional stepper motor that rotates the container along the ϕ axis with steps of 1.8 deg, and repeating the electronic scanning at each of these steps, 216 samples are collected along the azimuthal direction. The OUTs are in the form of cylinders with a height of 50 mm and a diameter of 20 mm, placed at the cylindrical coordinates (r, ϕ, z) of (35 mm, -60 deg, 0) and (50 mm, 125 deg, 0). Figure 10a shows the reconstructed images of the two objects. A comparison of the two sets of experimental results in Figures 9 and 10 shows that the increase in the number of samples has a positive impact on the quality of the reconstructed images, helping to reduce the background artifacts significantly.

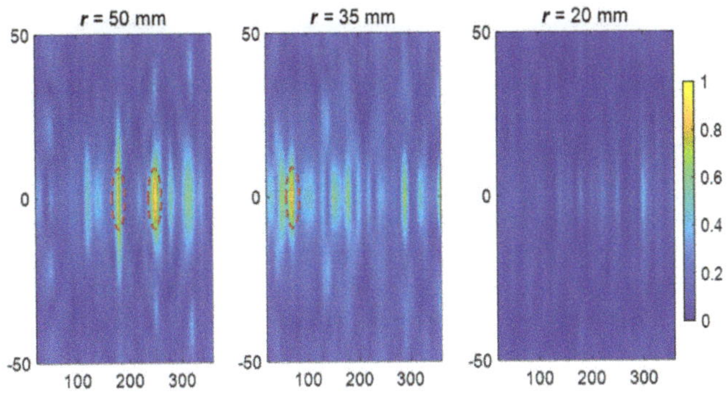

(a)

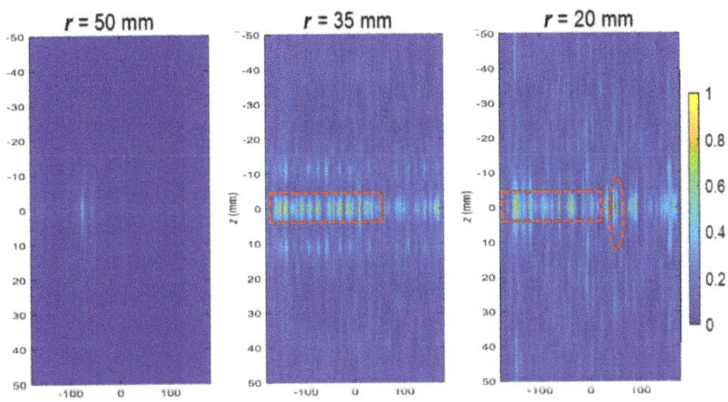

(b)

Figure 10. Measurement results for (**a**) Measurement Example 3 and (**b**) Measurement Example 4 when 216 samples along the ϕ axis are used. The red dashed lines show the positions of the objects. For all the figures, the horizontal axis represents ϕ (deg) and the vertical axis represents z (mm).

To further test the capability of this imaging approach, when using a combination of mechanical and electronic scanning, we perform imaging of a more complicated scenario (Measurement Example 4) in which a narrow and long arc-shaped object is placed on the middle imaged surface (at 36 mm) and along the azimuthal direction and another cuboid object similar to the ones used in the previous example is placed at (20 mm, 90 deg, 0). Figure 10b shows the imaging results for this example. It is observed that the objects are reconstructed in their true positions. However, larger shadows of the arc-shaped object are observed on the adjacent imaged plane at 20 mm. We believe that this degradation of the image quality is due to the fact that this imaging technique is based on the Born approximation [8] which provides better imaging results for smaller and lower-contrast objects (since it ignores multiple scattering between the objects).

5. Discussion

In this section, we provide important discussions related to the comparison of the proposed system with a previously proposed system, selection of the number of antennas and number of samples along the azimuthal direction, and distinguishing the objects in the reconstructed images.

5.1. Comparison of the Two Technique

In near-field microwave holographic imaging, the cross-range resolution of the reconstructed images depends on the portion of measurable evanescent waves, which, in turn, depends on the distance of the object and the antennas and the conductivity of the background medium. Thus, for imaged surfaces closer to the antennas, higher resolutions are expected. Since the distances between the antennas and the imaged surfaces and the conductivity of the background medium in this work are similar to those in [20], the azimuthal and longitudinal resolutions are similar to the ones reported in that work, i.e., 6 mm and 14 mm, respectively. Here, to demonstrate the performance of our imaging system in terms of resolution level, different positioning of objects in the imaged domain are shown in the presented examples. In general, enhancing the number of samples mainly reduces the shadowing level along the range direction.

Furthermore, the data acquisition time depends on the electronic switching time for selecting each pair of transmitter and receiver antennas t_{elec}, mechanical scanning step time t_{mech}, and measurement time of the scattered field t_{meas} (performed by VNA here and by customized data acquisition circuitry in [21]). Thus, in this work the total measurement time is $T_{\text{total}} = N_A \times t_{\text{elec}} \times t_{\text{meas}} \times 6t_{\text{mech}} \times N_z$, while in [21] it is $T_{\text{total}} = 8 \times t_{\text{elec}} \times t_{\text{meas}} \times 180t_{\text{mech}} \times N_z$. Since in general, the mechanical scanning time t_{mech} is much longer than the electronic scanning time t_{elec} due to the required delays in the system to stabilize mechanically, the data acquisition time here is much shorter than the system in [21]. By considering the parameters used in this work and [21], we expect the system to be approximately seven times faster in terms of data acquisition time.

Lastly, the system proposed in this work utilizes more antennas and switching modules to implement the discussed electronic switching scheme. While the system in [21] uses 9 antennas (1 transmitter and 8 receivers) and one SP8T switching module, the proposed system here uses 36 antennas and 6 SP8T switching modules. Thus, the implementation cost of the system here is higher.

Table 4 summarizes the comparison of the proposed system here with the one in [21] as discussed above.

Table 4. Comparison of mechanical and electro-mechanical scanning techniques.

Technique	Data Acquisition Time	Resolution	Cost	Shadow Effect
Mechanical Scanning [21]	$7T_{\text{total}}$	Same	Lower	Higher
Mechanical and Electronic Scanning (this work)	T_{total}	Same	Slightly Higher	Reduced

5.2. Selection of Number of Antennas and Azimuthal Samples for Experimental System

As discussed earlier in the article, the number of samples along the azimuthal direction (φ) is determined based on the physical size of the antennas to uniformly cover a full circle (360 deg). In this work, for dense sampling (as per physical constraints) and to maintain a compact array suitable for laboratory setting, an angular spacing of (10 deg) was selected, resulting in (360 deg/10 deg = 36) antennas distributed uniformly around the aperture. Due to space constraints and taking into account the cost of the system (number of utilized antennas and switching networks), it is not favorable to increase the number of antennas further. Thus, to enhance angular sampling, we employed a hybrid approach that integrates mechanical rotation steps along the φ direction with electronic scanning. Specifically, here the antenna array is mechanically rotated in six incremental steps, each of $10^\circ/6 = 1.7^\circ$, and at each step, the full electronic scanning sequence is repeated. The number of mechanical scanning steps was chosen as a compromise between the data acquisition time and quality of the reconstructed images. Consequently, the total number of azimuthal samples becomes 36 (electronic steps) $\times$ 6 (mechanical positions) = 216 at each height, significantly improving angular sampling without increasing the antenna count.

5.3. Distinguishing Reconstructed Objects from Artifacts

To distinguish between reconstructed objects and imaging artifacts, we analyze the brightness levels and spatial consistency across adjacent imaging planes. Genuine objects typically appear with higher brightness intensities and often produce weaker but noticeable shadows on neighboring planes, indicating their volumetric presence. In contrast, artifacts usually exhibit lower brightness and lack continuity or corresponding shadows in adjacent planes, making them visually separable from true object features.

6. Conclusions

In this paper, we introduced a novel near-field holographic microwave imaging system designed to overcome the long data acquisition time of the technique proposed in [21]. By integrating an electronically controlled switching mechanism, the proposed system enables virtual rotation of the antenna array along the azimuthal direction. This significantly reduces the data acquisition time along the azimuthal direction to barely 2 min long at each scanned height. Here, the data collection hardware is implemented in such a way that a space-invariant system is constructed (which is necessary to develop the convolution-based near-field holographic imaging technique), leading to enhanced imaging results. Thus, complex imaging scenarios in which the objects are at multiple range positions can be reconstructed satisfactorily. The methodology was validated through simulations and experiments, demonstrating the system's capability to accurately reconstruct objects' images at multiple range positions with enhanced resolution and reduced artifacts. Overall, the system successfully minimizes the common mechanical scanning limitations faced earlier in the systems relying solely on mechanical scanning along the azimuthal direction, such as slow data acquisition time, undesired measurement ripples, unwanted scanning vibrations, etc. This makes it a promising option for various applications demanding fast imaging.

Author Contributions: Conceptualization, R.K.A.; methodology, R.K.A. and M.H.; software, R.K.A. and M.H.; validation, R.K.A. and M.H.; resources, R.K.A.; data curation, M.H.; writing—original draft preparation, M.H.; writing—review and editing, R.K.A. and M.H.; supervision, R.K.A.; project administration, R.K.A. All authors have read and agreed to the published version of the manuscript.

Funding: This research received no external funding.

Data Availability Statement: Data are contained within the article. The data presented in this study can be requested from the authors.

Conflicts of Interest: The authors declare no conflicts of interest.

References

1. Zoughi, R. *Microwave Non-Destructive Testing and Evaluation*; Kluwer: Norwell, MA, USA, 2000.
2. Ghattas, A.; Al-Sharawi, R.; Zakaria, A.; Qaddoumi, N. Detecting defects in materials using nondestructive microwave testing techniques: A comprehensive review. *Appl. Sci.* **2025**, *15*, 3274. [CrossRef]
3. Nikolova, N.K. Microwave near-field imaging of human tissue and its applications to breast cancer detection. *IEEE Microw. Mag.* **2011**, *12*, 78–94. [CrossRef]
4. Origlia, C.; Rodriguez-Duarte, D.O.; Tobon Vasquez, J.A.; Bolomey, J.-C.; Vipiana, F. Review of Microwave near-field sensing and imaging devices in medical applications. *Sensors* **2024**, *24*, 4515. [CrossRef] [PubMed]
5. Sheen, D.M.; McMakin, D.L.; Hall, T.E. Three-dimensional millimeter-wave imaging for concealed weapon detection. *IEEE Trans. Microw. Theory Techn.* **2001**, *49*, 1581–1592. [CrossRef]
6. Ahmed, S.S. Microwave imaging in security—Two decades of innovation. *IEEE J. Microw.* **2021**, *1*, 191–201. [CrossRef]
7. Sheen, D.M.; McMakin, D.L.; Hall, T.E. Near-field three-dimensional radar imaging techniques and applications. *Appl. Opt.* **2010**, *49*, E83–E93. [CrossRef] [PubMed]

8. Amineh, R.K.; Nikolova, N.K. Fourier-space image reconstruction: The path toward real-time microwave and millimeterwave imaging. *IEEE Microw. Mag.* **2024**, *25*, 36–56. [CrossRef]
9. Tajik, D.; Pitcher, A.D.; Nikolova, N.K. Comparative study of the Rytov and Born approximations in quantitative microwave holography. *Prog. Electromag. Res. B* **2017**, *79*, 1–19. [CrossRef]
10. Tajik, D.; Kazemivala, R.; Nikolova, N.K. Real-time imaging with simultaneous use of Born and Rytov approximations in quantitative microwave holography. *IEEE Trans. Microw. Theory Tech.* **2022**, *70*, 1896–1909. [CrossRef]
11. Zhang, Z.; Liu, H.; Gao, X.; Zhang, Z.; He, Z.S.; Liu, L.; Zong, R.; Qin, Z. Two-Step Iterative Medical Microwave Tomography. *Sensors* **2024**, *24*, 6897. [CrossRef]
12. Slaney, M.; Kak, A.C.; Larsen, L.E. Limitation of imaging with first-order diffraction tomography. *IEEE Trans. Microw. Theory Tech.* **1984**, 32, 860–874. [CrossRef]
13. Chew, W.C. *Waves and Fields in Inhomogeneous Media*; IEEE Press: New York, NY, USA, 1990.
14. Wang, Y.M.; Chew, W.C. An iterative solution of two-dimensional electromagnetic inverse scattering problem. *Int. J. Imaging Syst. Technol.* **1989**, *1*, 100–108. [CrossRef]
15. Chew, W.C.; Wang, Y.M. Reconstruction of two-dimensional permittivity distribution using the distorted Born iterative method. *IEEE Trans. Med. Imaging* **1990**, *9*, 218–225. [CrossRef]
16. Zaeytijd, J.D.; Franchois, A.; Eyraud, C.; Geffrin, J.M. Full-wave three-dimensional microwave imaging with a regularized Gauss–Newton method—Theory and experiment. *IEEE Trans. Antennas Propag.* **2007**, *55*, 3279–3292. [CrossRef]
17. Van den Berg, P.M.; Kleinman, R.E. A contrast source inversion method. *Inverse Probl.* **1997**, *13*, 1607–1620. [CrossRef]
18. Chen, X.; Wei, Z.; Li, M.; Rocca, P. A review of deep learning approaches for inverse scattering problems (invited review). *Prog. Electromagn. Res.* **2020**, *167*, 67–81. [CrossRef]
19. Amineh, R.K.; McCombe, J.; Khalatpour, A.; Nikolova, N.K. Microwave holography using point-spread functions measured with calibration objects. *IEEE Trans. Instrum. Meas.* **2015**, *64*, 403–417. [CrossRef]
20. Amineh, R.K.; Ravan, M.; Sharma, R.; Baua, S. Three-dimensional holographic imaging using single frequency microwave data. *Int. J. Antennas Propag.* **2018**, *2018*, 6542518. [CrossRef]
21. Wu, H.; Amineh, R.K. A low-cost and compact three-dimensional microwave holographic imaging system. *Electronics* **2019**, *8*, 1036. [CrossRef]
22. Ghasr, M.T.; Horst, M.J.; Dvorsky, M.R.; Zoughi, R. Wideband microwave camera for real-time 3-D imaging. *IEEE Trans. Antennas Propag.* **2017**, *65*, 258–268. [CrossRef]
23. Sheen, D.; McMakin, D.; Hall, T.E. Cylindrical millimeter-wave imaging technique for concealed weapon detection. In Proceedings of the 26th AIPR Workshop: Exploiting New Image Sources and Sensors, Washington, DC, USA, 15–17 October 1997; Volume 3240, pp. 242–250.
24. Sheen, D.M. Sparse multi-static arrays for near-field millimeter-wave imaging. In Proceedings of the 2013 IEEE Global Conference on Signal and Information Processing, Austin, TX, USA, 3–5 December 2013; pp. 699–702.
25. Wu, H.; Ravan, M.; Amineh, R.K. Holographic near-field microwave imaging with antenna arrays in a cylindrical setup. *IEEE Trans. Microw. Theory Tech.* **2021**, *69*, 418–430. [CrossRef]
26. Manolakis, D.G.; Ingle, V.K. *Applied Digital Signal Processing: Theory and Practice*; Cambridge University Press: Cambridge, UK, 2011.
27. Pascual-Marqui, R.D. Standardized low-resolution brain electromagnetic tomography (sLORETA): Technical details. *Methods Find. Exp. Clin. Pharmacol.* **2002**, *24*, 5–12. [PubMed]
28. Altair Feko. Available online: https://altair.com/feko (accessed on 3 March 2025).
29. Golnabi, A.H.; Meaney, P.M.; Geimer, S.D.; Paulsen, K.D. Comparison of no-prior and soft-prior regularization in biomedical microwave imaging. *J. Med. Phys.* **2011**, *36*, 125–136. [CrossRef] [PubMed]
30. Golnabi, A.H.; Meaney, P.M.; Paulsen, K.D. Tomographic microwave imaging with incorporated prior spatial information. *IEEE Trans. Microw. Theory Tech.* **2013**, *61*, 2129–2136. [CrossRef]
31. Zhou, W.; Bovik, A.C.; Sheikh, H.R.; Simoncelli, E.P. Image quality assessment: From error visibility to structural similarity. *IEEE Trans. Image Process.* **2004**, *13*, 600–612. [CrossRef]
32. Adafruit Product 1858. Available online: https://www.adafruit.com/product/1858 (accessed on 3 March 2025).
33. Meaney, P.M.; Fox, C.J.; Geimer, S.D.; Paulsen, K.D. Electrical characterization of glycerin: Water mixtures: Implications for use as a coupling medium in microwave tomography. *IEEE Trans. Microw. Theory Techn.* **2017**, *65*, 1471–1478. [CrossRef]

MDPI AG
Grosspeteranlage 5
4052 Basel
Switzerland
Tel.: +41 61 683 77 34

Electronics Editorial Office
E-mail: electronics@mdpi.com
www.mdpi.com/journal/electronics